Palgrave Studies in the Theory and History of Psychology

Series Editor
Thomas Teo, Department of Psychology, York University, Toronto, ON, Canada

Palgrave Studies in the Theory and History of Psychology publishes scholarly books that draw on critical histories and theoretical concepts and methods, from a variety of approaches in the psychological humanities, to examine the discipline, profession, and practice of psychology.

This series publishes scholarly books that use historical and theoretical methods to critically examine the historical development and contemporary status of psychological concepts, methods, research, theories, and interventions. Books in this series are characterised by one, or a combination of, the following: (a) an emphasis on the concrete particulars of psychologists' scientific and professional practices, together with a critical examination of the assumptions that attend their use; (b) expanding the horizon of the discipline to include more interdisciplinary and transdisciplinary work performed by researchers and practitioners inside and outside of the discipline, increasing the knowledge created by the psychological humanities; (c) "doing justice" to the persons, communities, marginalized and oppressed people, or to academic ideas such as science or objectivity, or to critical concepts such social justice, resistance, agency, power, and democratic research. These examinations are anchored in clear, accessible descriptions of what psychologists do and believe about their activities. All the books in the series share the aim of advancing the scientific and professional practices of psychology and psychologists, even as they offer probing and detailed questioning and critical reconstructions of these practices. The series welcomes proposals for edited and authored works, in the form of full-length or short monographs; contact beth. farrow@palgrave.com for further information.

Series Editor:
Thomas Teo is Professor of Psychology at York University, Canada

Series Editorial Board:
Lisa M. Osbeck, University of West Georgia, USA
Annette Mülberger, University of Groningen, The Netherlands
Alex Gillespie, London School of Economics and Political Science, UK
Alexandra Rutherford, York University, Canada
Suzanne R. Kirschner, College of the Holy Cross, USA
Ernst Schraube, Roskilde University, Denmark
Antonia Larrain, Universidad Alberto Hurtado, Chile
Wahbie Long, University of Cape Town, South Africa

William H. Tucker

'The Bell Curve' in Perspective

Race, Meritocracy, Inequality and Politics

William H. Tucker
Rutgers, The State University
of New Jersey
Camden, NJ, USA

ISSN 2946-2452   ISSN 2946-2460   (electronic)
Palgrave Studies in the Theory and History of Psychology
ISBN 978-3-031-41613-2   ISBN 978-3-031-41614-9   (eBook)
https://doi.org/10.1007/978-3-031-41614-9

© The Editor(s) (if applicable) and The Author(s) 2024. This book is an open access publication.

Open Access This book is licensed under the terms of the Creative Commons Attribution 4.0 International License (http://creativecommons.org/licenses/by/4.0/), which permits use, sharing, adaptation, distribution and reproduction in any medium or format, as long as you give appropriate credit to the original author(s) and the source, provide a link to the Creative Commons license and indicate if changes were made.
The images or other third party material in this book are included in the book's Creative Commons license, unless indicated otherwise in a credit line to the material. If material is not included in the book's Creative Commons license and your intended use is not permitted by statutory regulation or exceeds the permitted use, you will need to obtain permission directly from the copyright holder.
The use of general descriptive names, registered names, trademarks, service marks, etc. in this publication does not imply, even in the absence of a specific statement, that such names are exempt from the relevant protective laws and regulations and therefore free for general use.
The publisher, the authors, and the editors are safe to assume that the advice and information in this book are believed to be true and accurate at the date of publication. Neither the publisher nor the authors or the editors give a warranty, expressed or implied, with respect to the material contained herein or for any errors or omissions that may have been made. The publisher remains neutral with regard to jurisdictional claims in published maps and institutional affiliations.

Cover illustration: © Kanate Chainapong/Alamy Stock Vector

This Palgrave Macmillan imprint is published by the registered company Springer Nature Switzerland AG
The registered company address is: Gewerbestrasse 11, 6330 Cham, Switzerland

Paper in this product is recyclable.

William H. Tucker: Deceased.

Foreword

Why should *The Bell Curve*, published nearly three decades ago, still deserve our attention and concern? Herrnstein and Murray's 1994 book was thoroughly critiqued by scholars for its deficient science, unjustified conclusions, misrepresentations of heredity, inappropriate source material, and more. For many social scientists, *The Bell Curve* was debunked and discredited, and could now be ignored. Richard Herrnstein had died before publication and could not defend against the accurate charge that ideas of an ineluctable hereditary racial hierarchy were at the heart of the book. But Charles Murray rejected all critiques of hereditary inequality as left-wing ideology or "Orwellian disinformation" as he called it in *Commentary* magazine in 2007. He has continued to promote a vision of permanent inequality, further weaving these ideas into the libertarian political ideology and warnings of the "decline of the West" that were central to his work even before *The Bell Curve*.

In this ongoing campaign, Murray was able to publish in popular magazines, *The Wall Street Journal, The Washington Post*, academic journals such as the psychology journal *Intelligence*, and numerous books, while simultaneously proclaiming that the topic was under a strictly enforced "taboo." With the backing of the American Enterprise Institute, his work was featured on conservative websites, podcasts, and talk shows, and *The Bell Curve* remained influential in White Nationalist circles. At the same time, psychologists continued to produce the discredited race science that Herrnstein and Murray had relied on, especially in publications by Richard Lynn and his younger and very active co-workers. Even after the 2012 death of two major contributors to scientific racism, psychologists Philippe Rushton and Arthur Jensen, hundreds of new articles, chapters, and books appeared that claimed scientific justification for permanent hereditary inequality.

William H. Tucker, Professor Emeritus of Psychology at Rutgers University-Camden, began his project of documenting the history and use of psychological science for oppression with his award-winning 1994 book, *The Science and Politics of Racial Research*. In two more books and a series of articles and chapters, he continued to show how this enterprise had been organized, funded, and sustained. Tucker provided an invaluable body of careful scholarship that is greatly admired by those who study the history of scientific racism.

This brief new work, *Race, Meritocracy, Inequality, and Politics: The Bell Curve in Perspective*, was completed but unpublished when Tucker died in 2022. Here he examines the implications of *The Bell Curve* for the social, economic, and political developments of the early twenty-first century. Following a review of the reception of *The Bell Curve* and its place in the campaign to end affirmative action, Tucker carefully examines Herrnstein's concept of the "meritocracy" in relation to earlier twentieth-century eugenics and the dramatic increase in economic inequality over the past 30 years. By examining the extreme rise in incomes in the fields of finance, corporate management, and law, Tucker shows how, contrary to *The Bell Curve*'s predictions, the reallocation of these huge sums was neither rational nor beneficial for society. The emergence of this "cognitive elite," heralded by Herrnstein and Murray as key to social progress and stability, eventually contributed to the damaging anti-elitism backlash of Donald Trump's populism. By demonstrating how *The Bell Curve* was clearly a social policy document that aimed to reshape society and eliminate social programs, Tucker provides an important new analysis of the interplay of science and politics, one informed by his lifelong commitment to social justice.

Andrew S. Winston
Professor Emeritus of Psychology
University of Guelph
Guelph, ON, Canada

Acknowledgements

Professor Tucker sadly died before publication could be completed. This work has been prepared for publication by his colleague Professor Daniel Hart with the permission of his widow Monica Drozd.

Contents

1 The *Bell Curve*, Then and Now   1
2 Meritocracy: Places, Everyone!   21
3 Politics and Intelligence: Running Against the Cognitive Elite   81
4 Conclusion: Addressing Inequality   101

Index   111

1

The *Bell Curve*, Then and Now

In the fall of 1994, the national obsession with the murder trial of a legendary football player was temporarily interrupted by a controversy over a book—not some sensationalized biography of a celebrity but a chart-filled 845-page tome, co-authored by a Harvard research psychologist and a policy wonk at the conservative think tank, the American Enterprise Institute. Despite its more than 100 pages of appendices on logistic regression and other technical, statistical issues, *The Bell Curve: Intelligence and Class Structure in American Life* became an instantaneous cause célèbre, providing its junior author, Charles Murray—Professor Richard J. Herrnstein having passed away only days before publication—with considerably more than his Warholian 15 minutes of fame and leading a reporter for the *New York Times Magazine* to designate him "the most dangerous" conservative in the country. Among the many "serious" periodicals to discuss the book at length, *The New Republic* devoted almost an entire issue to an essay by its authors along with a host of responses, and for some weeks Murray was a ubiquitous presence on television talk shows.[1]

The appearance of *The Bell Curve* was a carefully orchestrated political event. Departing from traditional procedure, the book was "embargoed" beforehand: no copies were circulated to potential reviewers or critics, a problem compounded by the fact that it was filled with the kind of statistical analyses normally published first in academic journals. Indeed, one prominent researcher maintained that none of the book's most important claims concerning a racial difference in IQ "could be published in any respectable peer-reviewed journal." Unsurprisingly *The Bell Curve*'s predominantly hereditarian explanation for the relationship between IQ and a host of variables such as income, welfare dependency, health, and quality of parenting elicited polarized reviews. Peter Brimelow—then an editor at the business magazine, *Forbes*, but soon to become a leader of the white nationalist movement—claimed, in a passive voice conveniently lacking an agent, that the book was being "seriously compared" with Darwin's *Origin of Species*, while a *New York Times* columnist described it as "a scabrous piece of racial pornography masquerading as serious scholarship," just "a genteel way of calling somebody a n***er" (the word was spelled out in the newspaper), and the award-winning American historian, Jacqueline Jones, called the book "hate literature with footnotes."[2]

More than two decades later, Murray's appearance on a college campus has continued to provoke controversy, and there have been a number of attempts—in the current term coined for the sort of treatment Murray has experienced—to "de-platform" him. His invited talk at Virginia Tech in spring 2016, part of a lecture series sponsored by a bank, elicited protests as well as a "counter-lecture," in which several professors participated in a teach-in denouncing Murray's work as "junk pseudoscience" and "racist"; while defending his right to speak, the university's president nevertheless declared that Murray's conclusions had been used "to justify fascism, racism and eugenics." Six months later at Yale, student protesters packed the hall where Murray's lecture was scheduled, and when he rose to begin, they stood up, announced a concurrent teach-in on the effects of white supremacy, and departed, leaving Murray to deliver his talk to an almost empty room. Before his spring 2017 appearance at Middlebury, one of the nation's premier liberal arts colleges, a number of faculty members signed a petition requesting that he be disinvited. Then when the talk took place, demonstrators shouted Murray down, chanting "Who is the enemy? White supremacy," and forcing him to move to another site to live stream his lecture while activists set off fire alarms; after the event, protesters pushed and shoved both Murray and the faculty moderator of the talk, who sustained whiplash and a concussion. And at the University of Michigan in October 2017, protester packed the hall and spent the better part of an hour chanting that Murray was a racist and projecting the words "white supremacist" on the wall, before marching out en masse.[3]

Also in spring 2017, in response to Murray's experience at Middlebury, the neuroscientist and outspoken atheist Sam Harris hosted Murray on his own podcast, during which the two men shared their dismay at the failure of so many people to accept what they considered the incontestable facts in *The Bell Curve*—the book's "Forbidden Knowledge" according to the conversation's self-congratulatory title. There was, Harris assured his audience, "virtually no scientific controversy" over Murray's argument, and as a result critics—not just the rowdy students but also those academics who expressed outrage over his physical harassment at Middlebury and voiced their disagreement in more professionally appropriate ways—could not possibly have acted in "legitimate good-faith" but were necessarily guilty of "dishonesty and hypocrisy and moral cowardice." Although none of Murray's critics were ever deplatformed, the accusation of ad hominem responses on their part was becoming little more than a tu quoque between the two sides. When the journalist Ezra Klein, editor of the news and opinion website Vox, subsequently published articles by three prominent research psychologists—two of whom held endowed chairs, one at the University of Virginia, the other at the University of Michigan—describing Murray's work as "junk science" and calling Harris "the latest to fall for it," Harris sent Klein a request to "stop publishing libelous articles about me," implying the possibility of legal action over a scientific dispute.[4]

The Role of Race

From the beginning of the controversy to the present, Murray has expressed surprise that *The Bell Curve* provoked such virulent opposition; like Captain Renault in *Casablanca* he has been shocked—shocked!— that his critics focused on the book's treatment of race. (After burning a cross as a high school student, he acknowledged being similarly "oblivious" to the possibility that the act had "any larger significance," though in retrospect he called his behavior "incredibly dumb.") Shortly after its publication, Murray attributed the reaction to "the American preoccupation with race." The book that his critics were discussing, he insisted, had precious little in common with the book he had actually written, one in which race played "a very small part ... tucked away in the middle." "In all," Murray pointed out, ethnic differences in intelligence only constituted "a major topic in four of the 22 chapters" and was alluded to peripherally in one of the two concluding chapters. More than 20 years later, he was making the same point in response to the statement by the Virginia Tech President. Charging that the President was "unfamiliar ... with the actual content of *The Bell Curve*," the topic of which was not race but class, Murray actually provided a count of the number of pages discussing genes, race, and IQ.[5] Yes, he seemed to be arguing, there was a turd in the punchbowl, but it was a really small one; why make such a big deal out of it?

Yet Murray himself had not been immune from the fixation he saw in others; indeed, he had a history of exploiting race to promote his publications. Murray's earlier book, *Losing Ground: American Social Policy, 1950–1980*, a highly controversial work that first brought him to public attention, urged the elimination of all federal antipoverty programs on the grounds that they had only exacerbated the problem of poverty by encouraging the poor to be lazy and irresponsible—again, an argument focused on class, in which race supposedly played but a marginal role. In his proposal to the publisher for *Losing Ground*, however, Murray pitched the work's sales potential in words that could also have applied to *The Bell Curve*: the book should sell well, he told the publisher, "because a huge number of well-meaning Whites fear that they are closet racists. And this book tells them that they are not. It's going to make them feel better about things they already think but do not know how to say." And in the earlier book itself Murray declared that "Real reform of American social policy is out of the question until we settle the race issue." In particular, he argued, the desire of white elites to make restitution for past sins of discrimination had led them to overlook or cover up numerous personal deficiencies on the part of poor Blacks that never would have been tolerated in Whites.[6]

Besides, notwithstanding the minor role he later claimed that race had played in *The Bell Curve*, at the time of its publication Murray did not hesitate to emphasize the book's discussion of race, once again as a way to generate publicity. Thus, when *The New Republic*, a nationally influential monthly focusing on politics and the arts, devoted its complete September 1994 issue to *The Bell Curve*, then on the verge of publication, Murray voiced no reservation over the magazine's cover, on which the word "RACE" occupied the entire top half of the page and the words "& I.Q." a substantial portion of the bottom half—an act guaranteeing that that topic would be the focus of attention. Indeed, Murray contributed the issue's featured 10,000-word essay, titled "Race, Genes and I.Q.–An Apologia," which began with the following candid statement:

> The private dialogue about race in America is far different from the public one, and we are not referring just to discussions among white rednecks. Our impression is that the private attitudes of white elites toward blacks is strained far beyond public acknowledgment, that hostility is not uncommon and that a key part of the strain is a growing suspicion that fundamental racial differences are implicated in the social and economic gap that continues to separate blacks and whites, especially alleged differences in intelligence.[7]

Only in hindsight did Murray apparently realize how insignificant a role race had played in the book.

In addition to dismissing the importance of race in *The Bell Curve*, Murray also emphasized the anodyne nature of the book's conclusion to its discussion of genes and racial differences in IQ, regularly quoting in full the "crucial" paragraph from the book:

> If the reader is now convinced that either the genetic or environmental explanation has won out to the exclusion of the other, we have not done a sufficiently good job of presenting one side or the other. It seems highly likely to us that both genes and the environment have something to do with racial differences. What might the mix be? We are resolutely agnostic on that issue; as far as we can determine, the evidence does not yet justify an estimate.[8]

That is, Murray thought it "highly likely" that genes were one reason Blacks were less intelligent than Whites, though he was "resolutely agnostic" on the extent—whether their disadvantage was large or small. Although naturally such a conclusion allowed for the latter possibility, *The Bell Curve*'s style, which the Harvard Professor of Education Howard Gardner termed "scholarly brinkmanship," strongly implied the former; as Gardner accurately characterized it, the book's "authors come dangerously close to embracing the most extreme positions, yet in the end shy away from doing so," thus encouraging "the reader to draw the strongest conclusion, while allowing the authors to disavow this intention."[9]

Nevertheless, up to the present the "agnostic" paragraph has been offered ad nauseam by both Murray and his supporters as an indication of the inoffensiveness of the book's conclusion about racial differences in intelligence. As Sam Harris put it in 2018, Murray was not claiming that the differences were substantially genetic, only that "genes and environment both play a part"; this was a truism, bordering on the banal, Harris argued, "as safe an assumption in behavior genetics as can be made." At the same time, Andrew Sullivan—who as editor of *The New Republic* when *The Bell Curve* appeared, had presided over the magazine's edition focusing on the book—also insisted that Murray has "remained resolutely 'agnostic'" about racial differences, just as he had been when the book was published. Murray himself, in his 2016 response to the President of Virginia Tech, again reproduced the concluding paragraph as a supposed indication of his thinking. In fact, more than a decade earlier the pretense of agnosticism had been dropped in an article by Murray in the neoconservative magazine, *Commentary*. While the print version omitted footnotes, the online "fully annotated version" included a citation to evidence that Murray judged "consistent" with the conclusion that 50–80 percent of the difference in IQ between Blacks and Whites—that is, from 7.5 to 12 points of the 15-point difference—was genetic.[10] Although he rarely acknowledged it in print, Murray was more of a true believer on racial differences than an agnostic.

Indeed, Murray's next major publication, nine years after *The Bell Curve*, offered additional reason to doubt his claims of agnosticism. Though it hardly became the cause célèbre of his previous work, *Human Accomplishment: The Pursuit of Excellence in the Arts and Sciences, 800 BC to 1950* provided a statistical demonstration of the overwhelming superiority of Western culture. Using as an operational definition of eminence a combination of frequency of an individual's appearance in reference books—encyclopedias and biographical dictionaries—together with the amount of space in column inches devoted to his or her entry, Murray calculated an "index score" allowing him to identify 3869 "persons that matter" overall from a number of categories: six specific sciences, mathematics, medicine, technology, Western music, art, literature, and philosophy. Although there were separate inventories for some ethnicities for the last three categories—for example, Arabic, Chinese, Indian, Japanese, and Western literature—Murray emphasized that, with rare exception, the index scores across these ethnic groups were not comparable because the pools for the non-Western groups were so much smaller; the number of "significant figures" in Western literature, for example, was almost triple the number in the other four groups combined. This methodology led him to conclude that white males from "a few places in Europe" had been responsible for "far more intense levels of human accomplishment" than any other people. And if one added to this "European core" the contributions of white men from North America, this small group accounted for almost every meaningful accomplishment by the human species. Exactly one black person qualified for inclusion on one of these lists: Duke Ellington received an index score of 2 on a scale 1–100, ranking him tied for 269th (along with 110 other persons) in the category Western music. Of course, it was not until the latter half of the twentieth century that Blacks enjoyed any access to the cultural mainstream, which did not prevent them, under systematically oppressive conditions and with no institutional support, from developing gospel, blues, jazz, reggae, and soul among other genres, not to mention their contributions to contemporary literature. However, Murray finessed the issue by "cutting off the inventories at 1950" on the grounds that even "expert opinion" may be "mostly a matter of fashion and ... quite different a hundred years from now." Besides, Murray insisted, citing as an example Toni Morrison—recipient of, among others, the Pulitzer Prize, the American Book Award, the Nobel Prize in Literature, and the PEN/Saul Bellow Award for Achievement in American Fiction—"women and black writers" were now being recognized "out of all proportion to their merit, in order to promote equality."[11] Not only did Blacks have a lower average IQ, but, according to Murray, in comparison with Whites they had produced almost nothing of enduring cultural significance.

Of course, whatever the length of *The Bell Curve*'s discussion of racial differences and whatever its conclusion, it was guaranteed to provoke a firestorm given the book's larger argument: if, as Herrnstein and Murray contended, socioeconomic status was in large part an (appropriate) reflection of genetically based intelligence, then their claim that Blacks were to some degree genetically disadvantaged could easily be exploited to justify the effects of discrimination as merely the inevitable consequence of biological differences. As Herrnstein had argued a few years before publication of *The Bell Curve*, there were "two fundamentally different models" to account for differences between Blacks and Whites in income and other variables related to the quality of life: the "discrimination" model, which attributed such differences largely to institutional policies that systematically disadvantaged Blacks; and the "distribution" model, which explained them as "the product of differing average endowments of people in the two races." In the latter view Blacks' economic status reflected, not their lack of opportunity but their lack of ability, making their lesser resources the economic reflection of their genetic merit. Indeed, according to *The Bell Curve*, after controlling for IQ the difference in annual wages between black and white workers almost entirely disappeared. If Blacks were thus clustered at the economic nadir largely because of their genes, then the society could not only ignore all those unfounded complaints about discriminatory laws and inequitable treatment but also evade any accountability for the lengthy history of practices that, for more than a century, had steadily converted black work into white wealth: convict labor, which helped to build the new South through the slave labor of prisoners incarcerated for crimes only charged against Blacks; home buying on contract, in which, much like purchase of a car, after a substantial down payment a single missed monthly payment entitled the seller to immediate repossession, depriving the owner of any equity and allowing the same house to be sold multiple times; FHA redlining, which prevented Blacks from borrowing money to buy a house; the exclusion of most Blacks from the benefits of the New Deal; the use of legal chicanery to swindle Blacks out of land ownership; and numerous other discriminatory practices designed to enrich Whites while keeping Blacks in economic servitude.[12] And if the massive gap between Blacks and Whites in both income and wealth had little to do with this history of racial oppression but rather resulted from certain characteristics of Blacks that were innate and immutable, then there was no reason to waste resources on social or economic programs senselessly intended to alter outcomes rooted in biology.

Yet another reason that Murray's disavowal of the significance of race lacked credibility had to do with what one critic called *The Bell Curve*'s "tainted sources." The book's authors acknowledged the assistance and cited the published works of numerous people associated with the *Mankind Quarterly* and the Pioneer Fund: the former an obscure journal founded in 1961 by a combination of European social scientists sympathetic to Nazi "Rassenhygiene" and American academics supportive of segregation, who sought to oppose the *Brown* decision on the basis of Blacks' genetic inferiority; the latter the source of financial support for the *Quarterly* as well as for every other scientist in the last half century intent on proving Blacks intellectually deficient, and whose board of directors had planned and executed a series of campaigns during the 1960s to block the civil rights movement. At the time *The Bell Curve* was published, the *Mankind Quarterly* was under the control of a British anthropologist and recipient of generous support from the Pioneer Fund, who had earlier, under a pseudonym, edited a journal dedicated to the view that World War II had resulted from an attempt by the Jews to exterminate the German nation and then, again under various pseudonyms, published an argument five times in the *Mankind Quarterly* (and another four in other journals under his control), insisting in almost identical language each time that racial prejudice was a biological necessity, essential "to maintain the integrity of the gene pool" and that interracial relationships, especially among "heavily urbanized and intellectually distorted human beings" constituted a "perversion" of natural instincts, similar to caged animals attempting "to mate with animals of other breeds."[13]

Although *The Bell Curve* made no mention of the Pioneer Fund, in its brief summary of the history of immigration and intelligence testing in the early twentieth century, the book referred coyly to a "biologist who was especially concerned about keeping up the American level of intelligence by suitable immigration policies." In fact, as the context made clear, this unnamed scientist was H. H. Laughlin, Pioneer's first president and an ardent admirer of the Third Reich, who had supported a program to deport all Blacks to Africa, calling their presence "the worst thing that had ever happened to the ... United States," and rejected a similar policy for Jews only because he recognized the practical difficulties in its implementation; as the next best step he hoped at least "to prevent more of them from coming."[14]

The Bell Curve relied on one Pioneer-funded scientist in particular for expertise on IQ testing. Herrnstein and Murray acknowledged having "benefited especially from the advice of Richard Lynn," whom they then describe as "a leading scholar of racial and ethnic differences." Indeed, as the Wellesley College historian Quinn Slobodian has discovered, Herrnstein's recently opened archives reveal that while working on the book the Harvard psychologist and his co-author had been sending drafts of chapters to Lynn for comment. At the time Lynn was an associate editor of the *Mankind Quarterly*, in which he had argued that, as a result of evolutionary selection, "Negroids" were considerably less intelligent than other races; indeed, according to Lynn, as low as their intelligence was, American Negroids still ranked considerably higher than their African counterparts, who had lacked the genetic benefits of "hybridization" with "Caucasoids." Lynn went on to become the leading scientific authority for white supremacists: editor of the *Mankind Quarterly* as well as, first a member of Pioneer's board of directors and then president of the fund; much of his work in the last two decades has appeared under the imprint of Washington Summit Publishers, owned and directed by

Richard Spencer, arguably the most prominent neo-Nazi in the United States.[15]

Another long-time recipient of financial support from the Pioneer Fund at the time of *The Bell Curve*'s appearance, J. Philippe Rushton—also later to become its president—proposed the application of an established biological theory concerning differences in reproductive strategy across species to differences between races. Some species produce large numbers of offspring per individual but invest little parental time or attention in their development, while other species produce few offspring but invest substantial effort in raising them. According to Rushton, human races could be placed on the same spectrum, with Blacks at the former end and Whites and Asians at the latter—a result of evolutionary differences that were also correlated with a constellation of variables including intelligence, brain size, social cohesion, infant mortality, altruism, law abidingness, mental health, and impulse control; Blacks in each case tended to score at the less advantageous end of the scale. Seemingly obsessed with sexuality, Rushton also claimed to find that Blacks had larger sexual characteristics: breasts and buttocks in women; penis size, both length and circumference, in men, who also ejaculated a greater distance than other races, according to questionnaire responses from participants at a local mall. As a result, he concluded, "It's a trade-off: more brains or more penis. You can't have everything." Thus the image of Blacks, or "Negroids," as Rushton called them: systematically less intelligent, more criminal, and sexually licentious, producing more offspring, whom they then tended to neglect. Though *The Bell Curve* conceded that this theory had not yet been fully confirmed, Herrnstein and Murray assured readers that Rushton, a frequent speaker at the convention of the white supremacist American Renaissance—itself a Pioneer grantee—was not "a crackpot or a bigot"; he had made a strong case, offering "increasingly detailed and convincing empirical reports of the race differences in some of the traits on his list" and citing "preeminent biological authority for his use of the concept of reproductive strategies."[16]

The point here is not to focus on *The Bell Curve*'s regard for Lynn's conclusions or Rushton's odious theory, but rather on Murray's bizarre puzzlement that, after citing such sources and pronouncements, others should not understand how little his book had to do with race. Indeed, his comment to an interviewer shortly after publication that "Some of the things we read to do this work, we literally hide when we're on planes and trains"[17] suggested that his dismay was disingenuous; there was little reason to hide studies of the relation between cognitive ability and socioeconomic success for the population in general.

The Bell Curve's authors themselves did not seek support from Pioneer, although while they were working on the book, Herrnstein suggested to Murray that they could "in a pinch, ask the Pioneer Fund for help"—an indication that they were aware of Pioneer's focus and considered their own project a suitable fit. It turned out that they were right: after the book's appearance Harry Weyher, the fund's president, at the time, expressed his regret at not having had the opportunity to contribute; had Herrnstein requested support from the fund, Weyher would have provided it "at the drop of a hat."[18]

Nor was it only the book's critics who found racial differences to be a significant theme; a number of readers acknowledged its effect in their own political evolution toward far-right extremism. Profiled innocuously in the *New York Times* as "the Nazi sympathizer next door" who had gone from "leftist rock musician ... to fascist activist," one of the founders of the Traditionalist Worker Party attributed his "political awakening" to books by Pat Buchanan and Murray; a participant in the Charlottesville torchlight rally, the TWP calls for "an independent White ethno-state in North America," its citizenship "limited to White persons and White persons alone" under a "National Socialist government." An even more prominent activist, Nathan Damigo, who spent five years in prison for armed robbery before founding the white supremacist group Identity Evropa, cited *The Bell Curve* along with writing by David Duke and Jared Taylor, the founder and editor of *American Renaissance*—works he had read while incarcerated—as the major influences on his thinking.[19]

A final reason that *The Bell Curve*'s treatment of racial differences, however brief, became such a flashpoint had to do with the explicit, practical purpose for the topic's inclusion: to set the stage for a ferocious tirade against "the system of affirmative action, in education and the workplace alike," which was "leaking poison into the American soul." Murray stated frankly that so much space in the book had been devoted to this policy issue—more than for any other—because affirmative action was predicated on the explicit assumption that "all [ethnic] groups have equal distributions of cognitive abilities"; by disproving this supposed underlying assumption, Murray sought to make a persuasive case for abolishing the policy. Indeed, *The Bell Curve* candidly acknowledged that, since exact knowledge about the genetics of racial differences would have no effect on any decision about the treatment of individuals, no reason existed for pursuing the issue other than its implications for affirmative action.[20]

Much of *The Bell Curve*'s evidence for the detrimental effect of affirmative action, especially in employment, came from anecdotes and newspaper stories; as the authors noted, "Private complaints about the incompetent affirmative-action hire are much more common than scholarly examination of the issue." One detailed example of such egregious incompetence due to minority preference in hiring, supposedly compromising the performance of the Washington, DC Police Department, relied on two sources: a report by the journalist, Tucker Carlson, writing in *Policy Review*—at the time the flagship publication of the conservative Heritage Foundation—and a four-part investigative series in the *Washington Post*. Carlson's entire article mentioned the word "race" exactly once—only in order to report that the Washington Police Department "officially denies the use of affirmative action on the basis of race" and a number of other categories. However, he also noted that, according to the recruiting office, applicants "can obtain additional points" for claiming residency in the district, an advantage likely to "severely restrict the pool of white applicants," in Murray's view, since the white population in the capitol was concentrated in professional areas, "with no significant white working-class neighborhoods"; presumably only in the latter communities were parents likely to raise children with the desire to protect and serve.[21]

In any event, as evidence for the harmfulness of affirmative action, *The Bell Curve*'s summary of Carlson's article quoted its few second- or third-hand and particularly lurid, anecdotal observations: according to Carlson, the president of the police lodge had heard from an instructor at the academy that some recruits "could not read or write," and another instructor claimed to have seen "people diagnosed as borderline-retarded graduate from the police academy." Murray felt no need to mention any of the other factors also described by Carlson, especially the department's lack of mechanical and technological support. At any given moment more than half the patrol cars were out of service, and at the time of Carlson's article, 12 of 19 cars in one of the most violent districts in the city were inoperable. In 1993, when electronic recordkeeping had become standard in most public facilities, only a few of the city's police offices had computers, and many even lacked typewriters, so that reports had to be handwritten, a method that often creates problems even (especially?) when the writer has a high IQ. Almost all the phones in police stations were rotary, and as a consequence unable to be equipped with voice mail, meaning that calls went unanswered after 5 PM when the clerical staff ended its work day.[22] Most people would probably believe that such primitive levels of support, along with a number of other organizational problems noted by Carlson, might have had some relevance to the decline in police effectiveness. But by not even acknowledging their existence, Murray thus created the impression that any such decline was attributable entirely to the intellectual shortcomings of officers who would never have been hired absent the benefit of racial preference. No modern phones, no computers, not enough vehicles—not worth noting; affirmative action illiterates were the whole problem. Though also unmentioned in *The Bell Curve*, Carlson's article ended by describing the praise for the performance of Washington police in middle-class neighborhoods with organized anti-crime groups; apparently the same affirmative action recruits functioned quite effectively with citizen cooperation.

Murray's other source for the substandard intelligence of recent recruits as the cause of deterioration of police performance, a four-part investigative series in the *Washington Post*, examined in depth the criminal justice system's low rate of arrest, trial, and conviction in homicide cases—i.e., not just the role of the police department but the prosecutor's office and the court system as well. That part of the series concerned with the police concentrated almost exclusively on the department's elite homicide detective squad, a group hardly composed of the putatively illiterate recent hires. And the major problem within this squad, according to the report, was the crushing caseload, resulting in night shifts that, after completion of the necessary paperwork, would end only a few hours before the next one was scheduled to begin; one of the top detectives had requested a three-month sick leave for stress after his assignment had increased from ten cases a year to ten a month. The prosecutor's office was similarly overwhelmed, frequently forced to seek postponements due to scheduling conflicts from having to handle multiple cases at the same time; occasionally, a case would even be dismissed by a judge when the prosecution was still not ready to proceed after the defendant had spent more than a year in jail awaiting trial. Finally, the investigative series also noted the external factors contributing both to the homicide increase and the difficulty in prosecuting such offenses: the "proliferation of guns" and the ease of their availability, as well as the increase in gang- and drug-related crime and the consequent reluctance of witnesses to come forward.[23] If Murray's references to Carlson's article were deceptively selective, his assertion that the *Washington Post* series "confirmed" Murray's version of Carlson's account, implying as it did that the newspaper too had found the substandard intelligence of affirmative action hires the cause of the poor record, had no justification whatsoever; not a single sentence in the four days of multiple daily stories suggested such a conclusion.

However, merely on a quantitative basis—i.e., by the proportion of the book devoted to the topic—Murray was correct: *The Bell Curve* concentrated primarily on the relation of IQ to class, not race, though its conclusions were no less inflammatory. In fact, much of the book constituted an arranged and not always compatible marriage between Herrnstein's insistence that genetically based intelligence was the strongest single predictor of socioeconomic success and Murray's contention that the state should do as little as possible to assist the disadvantaged—a goal he has advocated in three other books, not to mention numerous articles.[24] But what contributed to the outrage over *The Bell Curve*, distinguishing it from these previous publications, all of which had emphasized cultural and environmental factors to justify the abolition of all types of official assistance to the poor, was its prediction of dystopian consequences for the society, should its policies fail to take biological differences in intelligence into account. Quite apart from any racial implications, the book offered ample reason for controversy. Indeed, its underlying rationale represented an extension of a similar argument made by Herrnstein two decades earlier in an article that itself had caused an uproar.

Notes

1. R.J. Herrnstein and C. Murray, *The Bell Curve: Intelligence and Class Structure in American Life* (New York: Free Press, 1994). J. DeParle, "Daring Research or 'Social Science Pornography'?: Charles Murray," *New York Times Magazine* (October 9, 1994): 50. See *New Republic* (October 31, 1994).
2. On withholding the book beforehand, see N. Lemann, "The Bell Curve Flattened," *Slate*, January 18, 1997, https://slate.com/news-and-politics/1997/01/the-bell-curve-flattened.html. R. Nisbett, "Race, IQ, and Scientism," in *The Bell Curve Wars: Race, Intelligence and the Future of America*, ed. S. Fraser (New York: Basic Books, 1995), 54. P. Brimelow, "For Whom the Bell Tolls," *Forbes* (October 24, 1994): 153. B. Herbert, "Throwing a Curve," *New York Times*, October 26, 1994, A27. J. Jones, "Back to the Future with *The Bell Curve*," in *The Bell Curve Wars*, ed. S. Fraser (New York: Basic Books, 1995), 93.
3. R. Korth, "Controversial Author Draws Protesters at Virginia Tech," *Richmond Times-Dispatch*, March 26, 2016. W. Bloom, W. Hilke, B. Hill, A. Jain, A. Lopez-Delgado, M. Menlo, and J. Meyers, "No, Law School Didn't Teach Us to 'Engage' with Racists," *Nation*, August 1, 2017. S. Saul, "Dozens of Middlebury Students Are Disciplined for Charles Murray Protest," *New York Times*, May 24, 2017. A. Stanger, "Understanding the Angry Mob at Middlebury That Gave Me a Concussion," *New York Times*, March 13, 2017. J. Arm, "We Brought Charles Murray to Campus. Guess What Happened," *New York Times*, October 12, 2017.
4. "Forbidden Knowledge—A Conversation with Charles Murray" can be accessed at https://samharris.org/podcasts/forbidden-knowledge/. E. Turkeimer, K.P. Harden and R.E. Nisbett, "Charles Murray Is Once Again Peddling Junk Science about Race and IQ: Podcaster and Author Sam Harris Is the Latest to Fall for It," *Vox*, May 18, 2017, https://www.vox.com/the-big-idea/2017/5/18/15655638/charles-murray-race-iq-sam-harris-science-free-speech. The "libelous" quote

appears in an email from Harris to Klein, published on the former's blog at https://samharris.org/ezra-klein-editor-chief/.
5. J. DeParle, "Daring Research or 'Social Science Pornography'?" 51–52. C. Murray, "The Real 'Bell Curve'," *Wall Street Journal*, December 2, 1994, A14. C. Murray, Booknotes Transcript (C-SPAN), December 4, 1994, 16. C. Murray, "An Open Letter to the Virginia Tech Community," American Enterprise Institute, March 17, 2016.
6. DeParle, "Daring Research or 'Social Science Pornography'?" 50. C. Murray, *Losing Ground: American Social Policy, 1950–1980* (New York: Basic Books, 1984), 221–223.
7. C. Murray and R.J. Herrnstein, "Race, Genes and I.Q.—An Apologia," *New Republic* (October 31, 1994): 27.
8. Herrnstein and Murray, *The Bell Curve*, 311.
9. H. Gardner, "Cracking Open the IQ Box," *The American Prospect* (Winter, 1995): 73.
10. See "The Sam Harris Debate" between Klein and Harris on *Vox*, April 9, 2018, https://www.vox.com/2018/4/9/17210248/sam-harris-ezra-klein-charles-murray-transcript-podcast. A. Sullivan, "Denying Genetics Isn't Shutting Down Racism, It's Fueling It," *New York Magazine*, March 30, 2018, http://nymag.com/daily/intelligencer/2018/03/denying-genetics-isnt-shutting-down-racism-its-fueling-it.html. C. Murray, "The Inequality Taboo," *Commentary* (September 2005): 13–22. See footnote 44 of the online version at https://www.commentarymagazine.com/production/files/murray0905.html.
11. C. Murray, *Human Accomplishment: The Pursuit of Excellence in the Arts and Sciences, 800 B.C. to 1950* (New York: HarperCollins, 2003): 81, 114, 295. Totaling up the "significant figures" in each category actually produces 4,002; the lower number is explained by people listed in more than one inventory. Murray's observation about women and black writers occurred in an interview with a British journalist about the book; see J. Coman, "White, Male, and Proud of It," *Telegraph*, January 31, 2004, https://www.telegraph.co.uk/culture/donotmigrate/3611273/White-male-and-proud-of-it.html.
12. R.J. Herrnstein, "Still an America Dilemma," *Public Interest* 98 (1990): 6. Herrnstein and Murray, *The Bell Curve*, 323. On convict labor: D.A. Blackmon, *Slavery by Another Name: The Re-Enslavement of Black Americans from the Civil War to World War II* (New York: Doubleday, 2008). On contract buying: B. Satter, *Family Properties: How the Struggle Over Race and Real Estate Transformed Chicago and Urban America* (New

York: Henry Holt, 2009). On Redlining: R. Rothstein, *The Color of Law: A Forgotten History of How Our Government Segregated America* (New York: Norton, 2017). On the New Deal: I. Katznelson, *When Affirmative Action Was White: An Untold History of Racial Inequality in Twentieth-Century America* (New York: Norton, 2005). On being cheated out of land: L. Presser, "The Dispossessed," *New Yorker* 95 (July 22, 2019): 28–35 and K. Holloway, "The Century-Long Fight," *Nation* (June 14–21, 2021): 8–9.

13. C. Lane, "The Tainted Sources of 'The Bell Curve'," *New York Review of Books* (December 1, 1994): 14–19. On the Pioneer Fund and *The Mankind Quarterly*, see W.H. Tucker, *The Funding of Scientific Racism: Wickliffe Draper and the Pioneer Fund* (Urbana: University of Illinois Press, 2002), especially 90–101, 159–179.
14. Herrnstein and Murray, *The Bell Curve*, 5. On Laughlin, see Tucker, *The Funding of Scientific Racism*, 44–48.
15. Herrnstein and Murray, *The Bell Curve*, xxv, 272. R. Lynn, "The Evolution of Racial Differences in Intelligence," *Mankind Quarterly* 32 (1991): 99–121. R. Lynn, "Race Differences in Intelligence: A Global Perspective," *Mankind Quarterly* 31 (1991): 255–296. Q. Slobodian, "Racial Science Against the Welfare State," unpublished paper presented at the History of Political Economy Workshop, Duke University, February 16, 2018. R. Lynn, *Race Differences in Intelligence: An Evolutionary Analysis* (Augusta, GA: Washington Summit, 2006/2015); R. Lynn, *The Global Bell Curve: Race, IQ and Inequality Worldwide* (Augusta, GA: Washington Summit, 2008); R. Lynn, *The Chosen People: A Study of Jewish Intelligence and Achievement* (Augusta, GA: Washington Summit, 2011); R. Lynn and T. Vanhanen, *IQ and Global Inequality* (Augusta, GA: Washington Summit, 2006).
16. J.P. Rushton, *Race, Evolution and Behavior: A Life History Perspective* (New Brunswick, New Jersey: Transaction Publishers, 1994). Rushton is quoted in A. Miller, "Professors of Hate: Academia's Dirty Secret," *Rolling Stone* (October 20, 1994): 110. Herrnstein and Murray, *The Bell Curve*, 642–643.
17. DeParle, "Daring Research or 'Social Science Pornography'?" 51.
18. Quoted in Slobodian, "Racial Science Against the Welfare State." Weyher is quoted in J. Sedgwick, "The Mentality Bunker," *GQ* (November 1994): 230.

19. R. Fausset, "In America's Heartland: the Voice of Hate Next Door," *New York Times*, November 26, 2017, A16. A. Beale and S. Kehrt, "Behind Berkeley's Semester of Hate," *New York Times*, August 4, 2017.
20. Herrnstein and Murray, *The Bell Curve*, 312–315, 479, 508.
21. Ibid., 492, 495–496.
22. T. Carlson, "D.C. Blues: The Rap Sheet on the Washington Police," *Policy Review* (Winter 1993): 26–73.
23. See A. Knight, "Of 1,286 Slaying Cases, 1 in 4 Ends in Conviction," *Washington Post*, October 24, 1993 A1, A19; A. Knight, "Deadly Hours," *Washington Post*, October 25, 1993, A1, A8; A. Knight, "When Clogged Courts Fail to Speed Justice," *Washington Post*, October 26, 1993, A1, A10; A. Knight, "Strategies to End the Carnage," *Washington Post*, October 27, 1993, A1, A16.
24. Murray, *Losing Ground*; C. Murray, *In Our Hands: A Plan to Replace the Welfare State* (Washington, DC: AEI Press, 2006); C. Murray, *Coming Apart: The State of White America, 1960–2010* (New York: Crown, 2012).

Open Access This chapter is licensed under the terms of the Creative Commons Attribution 4.0 International License (http://creativecommons.org/licenses/by/4.0/), which permits use, sharing, adaptation, distribution and reproduction in any medium or format, as long as you give appropriate credit to the original author(s) and the source, provide a link to the Creative Commons license and indicate if changes were made.

The images or other third party material in this chapter are included in the chapter's Creative Commons license, unless indicated otherwise in a credit line to the material. If material is not included in the chapter's Creative Commons license and your intended use is not permitted by statutory regulation or exceeds the permitted use, you will need to obtain permission directly from the copyright holder.

2

Meritocracy: Places, Everyone!

Although *The Bell Curve* represented Murray's first published discussion of genes and intelligence—as tactical support for his preferred policies ending assistance to the poor—his co-author had been writing on the topic for more than two decades. Herrnstein's initial interest in intelligence marked a radical departure from his previous work. As a Harvard graduate student in the early 1950s, he had studied with the famous behaviorist B.F. Skinner, specializing in operant conditioning with pigeons. Appointed to a junior faculty position at Harvard in 1958, he received promotion and tenure only three years later, after formulating the "matching law," an important theoretical result predicting that, when an organism is offered two response alternatives, the ratio between them will match the ratio of reinforcements associated with each alternative. His reputation well established as one of the world's experts on pigeon behavior, Herrnstein went on to occupy an endowed chair at Harvard.

In 1965 Herrnstein chose *The Atlantic Monthly* (soon to become *The Atlantic*), a mass periodical for educated laypersons, to share with the public the wondrous practical applications of his research expected to occur in the near future. Birds and other animals, he predicted, would eventually replace human labor in industry whenever a simple sensory task, as opposed to an exercise of judgment, was involved. He noted, for example, that pigeons could detect defective parts with greater accuracy than humans and could work longer with no sign of fatigue or the deterioration in standards displayed by their human counterparts. The full commercial exploitation of these capabilities, Herrnstein explained, was being "held back only by negative attitudes that oppose the dictates of good business"—perhaps an allusion to possible unemployment created by the use of animals. (In fact, when, in summer 2017, Western Michigan University rented a group of goats to clear some brush on campus, the local AFSCME chapter filed a grievance, contending that the goats were taking jobs from laid-off union members.) Herrnstein also hinted at important uses for pigeons in scanning reconnaissance photographs but was unable to provide details because of "security restrictions," a project probably of more significance during the Vietnam War.[1] However, these exciting applications failed to materialize, and in place of the glamor and enthusiasm once promised by the field of animal learning, it became a small backwater of experimental psychology.

Half a dozen years later Herrnstein returned to *The Atlantic*, though this time at the opposite end of the theoretical spectrum, no longer focusing on principles of learning and conditioning but now concerned with genes and intelligence. In an article titled simply "I.Q." Herrnstein sounded all the themes that would emerge in substantially greater detail two decades later in *The Bell Curve*; widely discussed and controversial, the article was soon expanded into a book. As he explained in the latter publication, after being "submerged for twenty years in the depths of environmentalistic [sic] behaviorism," Herrnstein's "confidence in the environmentalist doctrine" had finally broken down when his "study of the subject of intelligence testing (or more broadly mental testing) persuaded [him] that the facts about people point to the role of genes in human society." However, an additional factor in this dramatic change of perspective was the firestorm that had occurred two years earlier in response to an article in the *Harvard Educational Review* by Berkeley Professor of Education Arthur Jensen (also a Pioneer grantee), titled "How Much Can We Boost IQ and Scholastic Achievement?".

Published not long after Lyndon Johnson's War on Poverty had included educational resources on the grounds that an increase in cognitive abilities would better enable the children of the poor to improve their socioeconomic condition, the first sentence of Jensen's article dismissed any such hopes, bluntly declaring that "Compensatory education has been tried and it apparently has failed." Jensen went on to explain the reason for failure: these programs had been based on the inaccurate belief that the poor academic performance of minority children stemmed from "social, economic and educational deprivation and discrimination" and thus that they would benefit from the same kind of cultural enrichment and additional instruction in basic skills enjoyed by middle-class children. The real disadvantage for poor and minority children, he maintained, came not from their conditions but from their biology; they were just genetically less intelligent. Much of the remainder of this lengthy article—at 123 pages it consumed almost the entire issue of the journal—presented a discussion of the concept of heritability, a technical term from behavior genetics indicating what proportion of variation in a trait is associated with variation in genotypes, followed by the suggestion that, rather than material assistance Blacks would benefit most from eugenic measures to discourage their least intelligent elements from reproducing.[2] For a publication in an academic journal, Jensen's article produced an unprecedented degree of outrage. One social scientist accused him of having done "injury to children," and other well-known psychologists called his work "academic manure," "obscene," and "abominable."[3] Student activists organized against Jensen, urging boycotts of his classes, interrupting his lectures, and demanding that he be fired.

Herrnstein felt strongly about what he regarded as Jensen's mistreatment—and not just from the civil libertarian point of view that the Berkeley professor should not have endured such harassment merely for expressing his opinion; he also believed that Jensen had made a compelling case. Thus, along with reflecting his own newly developed interest in intelligence, the *Atlantic* article also served to provide intellectual support for Herrnstein's beleaguered disciplinary colleague, whose controversial *Harvard Educational Review* article he described as "cautious and detailed, far from extreme in position or tone." Although Herrnstein had never conducted any research on intelligence nor a fortiori on its genetic basis—never published anything on the topic in a professional journal—he was an effective popularizer. In readable prose for a mass audience, his *Atlantic* article described the development of the concept of intelligence, the measurement of which Herrnstein called "psychology's most telling accomplishment to date," and its importance in determining life outcomes. Turning to "the inherited factor in I.Q.," Herrnstein explained the meaning of heritability and described the most straightforward method for its estimation—the similarity between the IQs of identical twins raised in separate homes, pairs of individuals sharing identical genotypes but different environments—concluding that Jensen and "most of the other experts in the field" were right: "the genetic factor is worth about 80 percent and ... only 20 percent is left to everything else," a result he considered "psychology's best proved socially significant empirical finding." On the most inflammatory issue of a genetic component to racial differences, Herrnstein declared that "the case is simply not settled," but he certainly thought that an answer was possible and found it "irritating" for inquiry to be "shut off because someone thinks society is best left in ignorance."[4]

Little of this discussion was particularly controversial until, in the last two pages, Herrnstein described the effect of hereditary factors on "social standing," concluding that the society was heading toward a genetic caste system with a biologically superior upper class—essentially a genetic aristocracy. This effect was so clear, Herrnstein wrote, that he could express it in the form of a syllogism:

1. If differences in mental abilities are inherited, and
2. If success requires those abilities, and
3. If earning and prestige depend on success,
4. Then social standing (which reflects earning and prestige) will be based to some extent on inherited differences among people.

However, after the syllogism's modest conclusion Herrnstein went on to describe a future in which social standing was not just "to some extent" related to heritable traits, envisioning instead a biologically stratified society with little possibility for mobility in either direction. At the nadir of this genetic hierarchy Herrnstein foresaw "precipitated out of the mass of humanity a low capacity ... residue that may be unable to master the common occupations, cannot compete for success and achievement, and are most likely to be born to parents who have similarly failed"—a metaphor that Francis Galton, founder of the eugenics movement, also had in mind when he referred to the lower classes as "the residuum." As technological advancement created new jobs demanding, in Herrnstein's analysis, higher IQ's, these hereditary defectives would be the most adversely affected, so that in his future society "the tendency to be unemployed may run in the genes of a family about as certainly as bad teeth do now." At the other end of the socio-genetic ladder would lie a new aristocracy, a class with greater wealth, power, and privilege, but unlike aristocracies of the past, which, Herrnstein emphasized, "were probably not much superior biologically to the downtrodden," this new privileged class would be entitled to its prerogatives because "when people can freely take their natural level in society, the upper classes will, virtually by definition, have greater capacity than the lower."[5]

According to Herrnstein's analysis, this scenario was inevitable. Although he agreed with Jensen's estimate that the heritability of IQ was around 0.80—i.e., that 80 percent of the differences in IQ between people were associated with differences in their genes—Herrnstein emphasized that the exact value of this statistic was not necessary to his argument. The more that improvements occurred in the society—resulting in more equitable legal, social, and educational conditions—the more heritability would increase; when environmental differences were minimized, only genes remained to explain the differences in outcomes. Thus, he insisted, biological stratification was the direct and inevitable consequence of maximal equality of opportunity, because the removal of arbitrary barriers and unfair advantages would only increase the significance of genetic factors in both IQ and its correlate, socioeconomic success; the "successful realization of contemporary political goals," he insisted, would result in "the growth of a virtually hereditary meritocracy." This was Herrnstein's most important point, what he most wanted the public to recognize: that "their political goals are fighting the nature of the beast." The egalitarian objective of a more equitable distribution of society's resources was not only exposed as an impossible fantasy in this view, but those who pursued it by calling for equal opportunity, would only create, to their own dismay, an even greater separation between classes: "Actual social mobility is blocked by innate human differences," he explained, "after the social and legal impediments are removed."[6]

At its core Herrnstein's argument sought to remove questions of resource distribution generally considered to be moral or political decisions and present them instead as biologically ineluctable. This view placed humane aspirations for greater socioeconomic equality on a collision course with science; social systems intended to reduce inequality, whatever their name—economic democracy, democratic socialism, etc.—were thus proved to be hopeless. The fault was not in our stars but in our genes, and as long as equality of opportunity was guaranteed, no amount of tinkering with social organization could avoid the inevitable: biostratification. Inequality in the social order reflected inequality in the natural order.

Thus, in this view the society was heading inevitably toward what Herrnstein called a "meritocracy," a word that he acknowledged taking from the British sociologist Michael Young's novel, *The Rise of the Meritocracy*, describing how, well into the twenty-first century, the principles of genetics had combined with the measurement of intelligence to create an intergenerational ruling elite, whose membership was determined by test score; Young coined the portmanteau by joining the Greek suffix for "rule" or "authority" to the Latin "meritus" the past participle of the verb meaning "to earn" or "deserve." Herrnstein praised the book as a "prescient" account of what to expect, already "catching the attention of alert social scientists," apparently oblivious to the fact that the novel was intended as a withering satire of a dystopian future, in which an insufferably smug and arrogant ruling class, secure in its scientifically demonstrated superiority and lacking any sense of social responsibility since their position on top was due entirely to their own genes, presides over a lower class, the members of which—the "technicians"—are forced to recognize the truth of their inferiority, and the consequent fact that their position on the bottom is both inevitable and appropriate. Any sense of political community in such a society has been completely lost. Herrnstein converted "meritocracy" from an intended pejorative into a positive, even titling his own book based on the *Atlantic* article "I.Q. and the Meritocracy." (In 2001 Young complained that his neologism "has gone into general circulation, especially in the United States," pointing out that "the book was a satire meant to be a warning.")[7]

Although no one prior to Herrnstein had provided as detailed a genetic argument, the notion that some people possess inborn qualities justifying their superior position in society has roots as old as antiquity. Aristotle believed in government by hoi aristoi—"the best," those with exceptional natural ability—and maintained that, because of differences in the power of reason, "just as some are by nature free, so others are by nature slaves, and for these latter the condition of slavery is both just and beneficial." And in *The Republic* Plato described Socrates's "myth of the metals," in which a citizen's value to and position in the city are determined by which of three metals characterizes his soul: gold for those best fit to rule, silver for those who assist the rulers, and iron and bronze for those—farmers and craftsmen—whose place is to obey.[8] But not until the twentieth century did some philosophers and social scientists suggest the premise underlying Herrnstein's analysis: that only egalitarian societies would ensure that these innate characteristics suiting people to specific roles did in fact exercise such a determinative effect. In 1903 the Scottish philosopher, David G. Ritchie, anticipated Herrnstein's argument, declaring that "the result of ... equality of opportunity will clearly be the very reverse of equality of social condition," since "the abolition of legal restrictions on free competition allows the natural inequalities of human beings ... to assert themselves"; even under "a socialistic regime, which fell short of a complete communism," Ritchie expected these "inequalities of condition" to emerge. In 1923 the prominent sociologist F.H. Hankins, later to become president of the American Sociological Association, maintained that the whole purpose of equal opportunity, and especially education for all, was to serve as the "principal means whereby the natural aristocracy of the country can be discovered and trained for the superior responsibilities it is to fill." Acknowledging that the goal was a Platonic society, in which each person "is fitted into the social order at a level corresponding to his innate powers," Hankins foresaw "an enormous difference between those at the top and those at the bottom in social value, in power and in financial rewards"; the elimination of "artificial handicaps," he wrote, in a conclusion offering

Herrnstein's rationale as support for Aristotle's pronouncement, would reveal "those born to rule" and those "born to be ruled."[9]

However, whatever the role of equal opportunity, for many social scientists it was creation of the "mental test" that converted this notion of organized genetic determinism from sociological speculation to practical possibility. Ecstatic at the thought that a seemingly objective measure taking a mere 40 minutes to administer could furnish the basis for a Platonic paradise, early intelligence testers were eager to create a society in which each person could be assigned a genetically appropriate place. Lewis Madison Terman, for example—a member of the National Academy of Sciences, probably the most well-known educational psychologist in the first half of the twentieth century, and described by one historian as the scientist most "responsible for making the IQ a household word"—called for testing to begin in the earliest grades so that those children destined to be "the world's hewers of wood and drawers of water" could be removed from the usual curriculum and "segregated in special classes ... given instruction which is concrete and practical" in order to make them "efficient workers." For all other children Terman urged that "vocational guidance" should begin no later than fifth or sixth grade, directing each student toward an intellectually "compatible" occupation by comparing the IQ score with the minimum necessary for success in that field. Such a procedure, he explained, would not only avert "selection of a vocation ... requir[ing] a higher grade of ability than the individual possesses" but also ensure that bright students did not "waste" their abilities in an occupation requiring "mediocre intelligence." Any IQ score above 85 for a barber, for example, was "so much dead waste."[10]

Sir Cyril Burt, too, the internationally eminent British psychologist and first member of his profession to be knighted (whose research on the heritability of intelligence was posthumously exposed as worthless and probably fraudulent), believed that it was "the duty of the state through its school service" to provide a child "the education most appropriate to his powers, and ... to place him in the particular type of education for which nature has marked him out." Thus, Burt proposed that each child be classified according to test score into one of eight IQ ranges, each range then corresponding to an educational category leading to a specific set of vocational possibilities. In addition, Burt specified the portion of the population that was expected to fall into each classification: only the highest range, for example, encompassing a mere tenth of a percent of the population, would enjoy a university education leading to a career in the professions or to a "higher administrative" position, while roughly 11 percent would fall into the next two ranges channeling them into "higher grade schools" and eventually technical positions; the overwhelming bulk of the population in the lower ranges would receive the appropriate education for their destiny as workers, either "skilled," "semi-skilled," "unskilled," or "casual." Education would furnish what Burt called "the key ... to social efficiency ... a place for every man and every man in his place."[11]

While Terman, Burt, and other prominent psychologists emphasized the efficiency of using IQ scores to determine one's course in life, Charles Spearman—the British psychologist who first posited the notion of a general intelligence factor ("g") and pioneered the statistical process of factor analysis—offered an additional and even more grandiose justification: harmony. Not only would the measurement of intelligence ensure that "each can be given an appropriate education, and therefore a fitting place in the state–just that which he or she demonstrably deserves," Spearman predicted, but he was certain that, as a result of testing, "Class hatred, nourished upon preferences that are believed to be unmerited, would seem at last within reach of eradication; perfect justice is about to combine with maximum efficiency."[12] This last sentence is remarkable for its utter cluelessness about human nature, implying as it did that, after learning of their test scores, the poor would accept their unenviable station if not cheerfully, then at least without resentment toward their betters, knowing that it was merely the rational, social reflection of their genetic inadequacy; once the members of the lower class appreciated that their inferior position was not *un*merited, social harmony would prevail.

Michael Young's novel provided a much more realistic account of the likely reaction from actual human beings confronted with evidence of their mediocrity. In an unjust society, one lacking equal opportunity for advancement, "the workers," Young observed, "could altogether dissociate their own judgments of themselves from the judgment of society."

Those on the lower rungs of the social ladder could console themselves with the thought that, but for circumstances, their life would have turned out much different: "Had I a proper chance I would have shown the world," was their perspective. Thus, as Young put it, "Educational injustice enabled people to preserve their illusions, inequality of opportunity fostered the myth of human equality." But in a meritocracy, he explained, it becomes harder for the poor to bear their allotted position: "all persons, however humble, know that they have had every chance," and as a consequence, "if they have been labelled 'dunce' repeatedly they cannot any longer pretend. ... Are they not bound to recognize that they have an inferior status–not as in the past because they were denied opportunity; but because they *are* inferior? For the first time in human history the inferior man has no ready buttress for his self-regard," and the result, Young observed, was detrimental both to the individual and to the society: those "who have lost their self-respect are liable to lose their inner-vitality ... and may only too easily cease to be either good citizens or good technicians."[13]

At the time of the *Atlantic* article, Herrnstein's interest in intelligence and meritocracy was informed primarily by the efficiency principle. Though not as rigidly deterministic as his predecessors—more inclined to allow the intellectual demands of different vocations to create a natural sorting mechanism—Herrnstein was especially concerned with the tails at each end of the intelligence spectrum. Like Jensen, he too believed that the Great Society programs, designed to assist poor and minority children, had been a "failure," calling it "imperative that the government stop throwing its money down a bottomless hole." But unlike Jensen, Herrnstein emphasized educational resources as a zero-sum game, in which well-meaning efforts to provide low achieving students with conditions similar to those enjoyed by high achievers amounted to "withholding educational advantages from gifted people and lavishing them on the less well endowed." Thus, he argued, instead of compensatory education the vain attempt to reduce educational inequities threatened to create a system of "compensatory deprivation," which might "reduce individual differences, but do so at the expense of those who are fortunate enough to have been well endowed to begin with"—an approach that Herrnstein compared to "depriving healthier people of some part of their medical care and diverting it to the unhealthy." This sort of "selective deprivation," he concluded, was not only "unfair" to the genetically advantaged but "a waste our society can ill afford."[14]

The Bell Curve continued this focus on the importance of appropriate treatment for the extreme IQ scores, high and low. Both Terman and Burt had despaired of educating the duller students for any purpose other than to become unskilled labor. Herrnstein and Murray were even blunter: "People in the bottom quartile of intelligence," they declared, "are becoming not just increasingly expendable in economic terms; they will sometime in the not-too-distant future become a net drag ... unable to perform that function so basic to human dignity: putting more into the world than they take out." And as a result, they concluded, "For many people, there is nothing they can learn that will repay the cost of the teaching."[15] From a cost-effectiveness perspective, there was no sense even attempting to educate a substantial portion of the population.

While resources were thus supposedly being squandered on a misguided attempt to educate the lower tail of the IQ distribution, the top five percent—the group dubbed the "cognitive elite" by *The Bell Curve*—was being deprived of their appropriate share according to Herrnstein and Murray. Again as in Herrnstein's earlier work, *The Bell Curve* urged a reallocation, shifting federal aid from programs for the disadvantaged to programs for the gifted—an unsurprising recommendation given the book's opinion of the former initiative as futile and the latter as essential. The cognitive elite—or, as Herrnstein and Murray candidly termed them, "the people who count in business, law, politics and our universities"—by virtue of their genetic advantage would inevitably grow up "segregated from the rest of society," attending "the elite colleges," enjoying "successful careers," and "eventually lead[ing] the institutions of this country, no matter what." Thus destined for positions of power and influence, they needed "education of a particular kind," one not only with higher standards but emphasizing "how to think about their problems in complex, rigorous modes" and "bring to their thinking depth of judgment and, in the language of Aristotle, virtue." Yet their actual treatment in school, Herrnstein and Murray complained, represented the "one clear and enduring failure of contemporary American education." It was essential, they insisted, that this "natural aristocracy" be prepared for their genetically appropriate role to govern; such an emphasis on critical thinking, crucial to the formation of a polity capable of democratic self-governance, would nevertheless be unbefitting for the rest of the population, which presumably needed education of a more practical kind. (Similarly, one provision of "The Charlottesville Statement," the white supremacist manifesto issued by Richard Spencer, one of the leaders of the neo-Nazi "Unite the Right" rally at the University of Virginia in 2017, maintained that higher education was "only appropriate for a cognitive elite dedicated to truth" and "improper, even detrimental" for the great majority, for whom "practical education–trade schools and apprenticeships—should be ... the norm.") Transferring resources from the disadvantaged to the cognitive elite was thus necessary for "the welfare of the nation, including the welfare of the disadvantaged."[16]

But efficiency was not *The Bell Curve*'s only concern. Like Spearman, Herrnstein and Murray also believed that acknowledgment of the fact of genetic inequality and its ineluctable social consequences was essential if people were to "live together harmoniously despite fundamental individual differences." Not as naïve as Spearman, however, they hoped that the less intelligent would find it in their best interest to accept their biologically determined lot in life but feared that, as a result of their failure to do so, the society was heading in an unfortunate direction. A widespread, "egalitarian" political ideal had fostered illusory hopes for improvement in the abilities of the cognitive underclass and correspondingly unrealistic expectations about their place in society. In response, Herrnstein and Murray predicted, "Over the next few decades, it will become broadly accepted by the cognitive elite that the people we now refer to as the underclass are in that condition through no fault of their own but because of inherent shortcomings about which little can be done." To protect themselves from this low-IQ group, the cognitive elite—whose interests were increasingly converging with the affluent, producing "an unprecedented coalition of the smart and the rich"— would gravitate toward "a new kind of conservatism," one "along Latin American lines, where to be conservative has often meant doing whatever is necessary to preserve the mansions on the hills from the menace of the slums below." Thus, to keep the underclass "out from underfoot" the cognitive elite would implement a "custodial state ... a high tech and more lavish version of the Indian reservation for some substantial minority of the nation's population," making it "difficult to imagine the United States preserving its heritage of individualism, equal rights before the law, free people running their own lives."[17]

The only hope for avoiding this dismal scenario, according to Herrnstein and Murray, was to adopt a social policy informed by a "wiser tradition," one derived from the great political thinkers, who, for thousands of years, had appreciated that people differed from each other in fundamental and important ways, fitting them to play specific roles. This was true from ancient philosophers, in both the East and West, who understood that "society was to be ruled by the virtuous and wise few," to the nation's founding fathers–Jefferson, Madison, Adams—who believed not in a democracy allowing an equal voice to all but in a republic ruled by the "natural aristoi." The point to describing the views of these men, Herrnstein and Murray emphasized, was not to appeal "to their historical eminence, but to their wisdom. We think they were right." Indeed, they noted, the "main purpose of education," according to Jefferson, was "to prepare the natural aristocracy to govern"; the "people who count" had to be groomed for their appropriate role in the society.[18]

Thus, *The Bell Curve* concluded with a cautionary tale about the contemporary risks of failing to adopt a Platonic social model out of a reluctance to accept the importance of genetic differences in intelligence. Spearman had assumed that, faced with the objective evidence of their inferiority, the less well endowed would accept their station, knowing that it was warranted; Herrnstein and Murray feared the dire consequences of their reluctance to do so. Instead of vain and misguided attempts to overcome genetic disadvantage, the real need was to find "valued places" for everyone, especially those at the lower end of the intelligence spectrum. And the major obstacle to doing so, according to Herrnstein and Murray, was society's rules "that are congenial to people with high IQs and that make life more difficult for everyone else." What was needed, therefore, was a simplification, creating rules that were clear and comprehensible to "just about everybody who is not part of the cognitive and economic elites": less government regulation; swift administration of criminal justice in which trial and punishment follow arrest "within a matter of days or weeks"; the elimination of Head Start, compensatory education, affirmative action, and government assistance for low income women who bore children; and the limitation of parental rights only to married couples, so that an unmarried mother had no legal basis for demanding child support from the father.[19] More than 500 pages of statistical analysis of test scores thus culminated in a call for a set of policies that Murray had pursued long before he and Herrnstein had crunched their first set of IQ data, though now presented not as a political choice but a scientific necessity.

Economic Inequality: The Gradient of Gain

Writing in the early 1970s, Herrnstein made one prediction that turned out to be remarkably prescient: economic inequality would increase dramatically in the coming years. Two decades later in the *National Review* Murray too predicted that "the price for first-rate cognitive skills will skyrocket," producing an "American caste system," and *The Bell Curve* foresaw a similar trend, though by that time it was hardly surprising since inequality had become well entrenched as a feature of the economy. In contrast Herrnstein's earlier analysis occurred during a period when gains were still spread fairly equally across the economic spectrum. From the beginning of what economists have labeled the "Great Compression" in the 1940s to the end of the 1970s, a bar graph plotting change in income against economic quintile looks almost like a picket fence, with each quintile enjoying approximately the same percentage increase; during those three decades, the share of the nation's wealth held by the richest 1 percent fell from 48 percent of the total to just above 20 percent, as a combination of strong unions, progressive taxation, and social norms produced an unprecedented downward distribution of income, leading to probably the most generalized material prosperity in history. Indeed, writing around the same time as Herrnstein, the eminent sociologist Daniel Bell noted "the steady decrease in income disparity among persons, which he attributed to technological advance, implying that the trend could be expected to continue.[20]

Yet, as Herrnstein predicted, exactly the opposite occurred. From the end of the 1970s until the present, the economy has experienced the Great Divergence, again one of the largest redistributions of wealth ever, but this time upward. Of course, there was substantial overall growth during this period, but instead of the rising economic tide lifting all boats, only the luxury yachts rode the waves, while small craft found themselves stuck in shallow water. The same bar graph displaying income gain by quintile now looks like an irregular staircase beginning with a very small step up—the increase in income for the first quintile—and an increasingly larger step for each subsequent quintile; the larger the income, the larger the percentage increase. But if the very highest earners are broken out separately from the top quintile, they enjoyed such a substantial increase that the trendline over the income groups changes from linear with a large positive slope to dramatically exponential; the higher up the distribution, the much steeper the rise in income. Adjusted for inflation, at the beginning of 2019 the lower half of the income distribution had seen no increase in income since 1980, and the average hourly wage for the working class had actually declined at the same time that, according to the Berkeley economists Emmanuel Saez and Gabriel Zucman, "for the highest 0.1 percent of earners, incomes have grown more than 300 percent; for the top 0.01 percent incomes have grown by as much as 450 percent; and for the tippy-top 0.001 percent—the 2300 richest Americans—incomes have grown by more than 600 percent." As a result the proportional difference between the top 1 percent and the bottom 99 percent is replicated by the difference between the top 0.01 percent—the 1 percent of the 1 percent—and the top 0.99 percent and yet again by the difference between the "tippy-top" 0.001 percent and the top 0.009 percent. In a particularly striking indication of the disparity in growth, a journalist specializing in finance reported that from 1990 to 2000, for every additional dollar earned by the bottom 90 percent of taxpayers, those in the top 0.01 percent earned an additional 18,000; the same figure for the period between 1950 and 1970 had been 162 dollars. Concerned that such an "extreme concentration of wealth means an extreme concentration of economic and political power," Saez and Zucman concluded that "just as we have a climate crisis, we have an inequality crisis."[21]

Other standard measures of economic inequality led to the same conclusion. The distribution of wealth, for example—the net value of all a person's assets after factoring in debt—has become even more skewed than annual income. A 2014 study by the same two economists found that the wealthiest one tenth of a percent of the population accounted for as much of the country's wealth as the bottom 90 percent combined, and according to a report from the Institute for Policy Studies, at the end of 2017 the three wealthiest individuals in the United States—Bill Gates, Warren Buffett, and Jeff Bezos—owned more than the 160 million people in the bottom 50 percent of the American population combined; the wealth of just these three people also exceeded the total wealth, adjusted for inflation, of the entire Forbes 400 in 1982, the list's inaugural year. In 2019 the Gini coefficient for the United States—the most widely used index of a society's economic inequality—reached its highest level since tracking began and is now larger than any of the European nations in the Organization for Economic Cooperation and Development; compared to these nations the United States also has the largest percentage of working age people who live in poverty.[22] Once equal to that of other affluent democracies like Canada and Norway, income inequality in the United States now exceeds that of countries like India, Indonesia, Haiti, and Vietnam. By every conceivable measure, inequality in the United States is approaching levels not seen since before the Roosevelt administration—Teddy Roosevelt.

While the country has experienced extreme inequality before—during the Gilded Age in the late nineteenth century, for example—the present context is qualitatively different from the past in two related ways. As the economist Thomas Piketty points out in his landmark study of wealth concentration and distribution, *Capital in the Twenty-first Century*, earlier instances were characterized "by very high incomes from capital, especially inherited capital." But in the contemporary United States, Piketty observes, "the peak of the income hierarchy is dominated by very high incomes from labor rather than by inherited wealth"; in contrast to the traditional view that inequality is rooted in the conflict between capital and labor, the major economic divide in the society now stems from differences within the ranks of working people. "It is hardly surprising," Piketty notes drily, "that the winners in such a society would wish to describe ... [it] as 'hypermeritocratic,' and sometimes they succeed in convincing some of the losers."[23] In his insightful book, *The Meritocracy Trap*, the Yale law professor Daniel Markovitz dubs these people the "superordinate working class"—graduates of elite universities with "immense skill, won through rigorous training," whose jobs require "intense, competitive, and enormously productive industry," and who enjoy annual wages in 7, 8, 9, and occasionally even 10 figures. No longer an idle aristocracy leading a life of extravagant leisure, many of the super-rich now work, and work hard, for their colossal incomes (engendering a sense that they are not in any way parasites but have truly "earned" their money, which then fuels the resentment at any increase in tax rate); Markovitz calls them "today's Stakhanovites."[24]

In addition, as a result the main source of inequality has been relocated on the economic spectrum. In the past and especially during the post-war democratization of the economy, the middle class and the rich tended to converge, leaving the difference between the poor and everyone else as the society's major economic fault line. But the substantial decline in the kind of desperate poverty that once engaged humanitarian sensibilities together with enormous increases for the rich has changed the nature of inequality, making the difference between the superrich and everyone else the new inflection point. As Markovitz points out, a measure of inequality like the Gini coefficient has not changed for the bottom 90 percent of income distribution through the last half century; indeed, for the bottom 70 percent it has actually fallen. It is the dramatic increase in inequality for the top 5 percent—what Markovitz calls "the income gap between the merely rich and the exceptionally rich"—that alone accounts for the rise in inequality for the population overall.[25]

This extreme elongation of the upper end of the income spectrum is not just a matter of some people being richer than others. Hemingway's famous (though mythical) retort to Fitzgerald may have been accurate at the time, but today a tiny slice of the American population enjoys an entirely different life from that of their fellow citizens, far beyond the mere fact that "they have more money." Of course, there have always been rich and poor neighborhoods; in the early 1980s the cultural and literary historian Paul Fussell referred to the very rich as "the class in hiding" for their "estates where you can't see the house from the road."

But there were also common experiences that the richest people shared with the middle class. Now, however, as Nelson D. Schwartz documents in his aptly named book, *The Velvet Rope Economy*, those fortunate enough to have the financial resources "rarely come into contact with people from other walks of life." Whether it's private skyboxes at athletic events, helicopters to avoid road traffic, charter flights or private suites at the airport within steps of the plane to avoid the departure lounge altogether, concierge medical services that ensure immediate access to the best doctors and most recent advances (including the coronavirus vaccine as soon as it became available), private firefighting services that will arrive at a property and take preventive action when a fire that might eventually pose a threat is still a safe distance away, even special back stage passes at theme parks like Universal, Disneyland, or Disney World, the exceptionally wealthy can purchase an "E-Z Pass" in life, enabling them "to zip past the everyday obstacles the rest of us have to contend with."[26]

Though writing decades before the degree of inequality became so dramatic, neither Herrnstein, in his 1971 article, nor Herrnstein and Murray in *The Bell Curve* had any doubt about the correct interpretation of this trend. Indeed, Herrnstein not only predicted such extreme economic differences as inevitable in the developing knowledge economy, he viewed them as a rational allocation of the society's resources, assuming that they constituted an appropriate reflection of genetic differences in intelligence. For Herrnstein, the linkage between IQ score and income served an important societal purpose, channeling the most intelligent people into the socially more useful because intellectually more demanding occupations. "By directing its approval, admiration, and money towards certain occupations," he wrote, "society promotes their desirability," and "thereby expresses its recognition ... of the importance and scarcity of intellectual ability." And by ensuring that these positions enjoyed greater money, power, and social status, society provided a "gradient of gain" corresponding to "inborn ability," as persons with superior intelligence sought the greater rewards associated with work of greater social value. Such a sensible mechanism, in Herrnstein's opinion, allowed society to "husband ... its intellectual resources," preventing their waste on efforts of little importance.[27] The cognitive elite deserved much more money than everyone else as long as they pursued those professions that most required the application of their intelligence, and providing the former was the best way to ensure the latter.

Although he viewed heredity as the major determinant of intelligence, for Herrnstein whether superior intellects were in fact put to the social use entitling them to greater reward was dependent primarily on Skinnerian principles of reinforcement. Indeed, he argued, if society was so foolish as to invert the gradient of gain so that bakers and lumberjacks—occupations that Herrnstein had singled out as performed by persons with an average IQ below that of the population in general and, as a consequence, appropriately less prestigious and less well remunerated—"got the top salaries and the top social approval," then "soon thereafter, the scale of I.Q.'s would also invert" so that these newly desirable jobs now attracted people with the highest scores. As an inevitable result, according to Herrnstein, "the top I.Q.'s would once again capture the top of the social ladder."[28] Attaching different rewards to different occupations allowed society to direct the flow of talented labor toward the most socially beneficial positions.

As evidence for the linkage between intelligence and social contribution, Herrnstein presented the average IQ for a number of occupations, specifically citing accountants and public relations specialists as some of the highest scoring groups. Thus, at the time, the "brightest" people were attracted to professions that arguably spent much of their time either assisting the affluent to avoid their share of the tax burden or in attempts to confuse image with substance and generally deceive the public; Noam Chomsky suggested that service to wealth and power provided the true reason for their greater compensation. At the other end of the intelligence spectrum were the predominantly blue-collar workers—people whose jobs required them to get their hands dirty—and whose just deserts, according to Herrnstein, were "poverty, or at least, relative poverty as compared to our society's successful people." Upholsterers and stonemasons—occupations near the low end of the IQ spectrum on Herrnstein's list—might be skilled artisans, able to restore antiques or create complex shapes out of rough pieces of rock, but as "technological advance changes the marketplace for I.Q.," he wrote, the new positions would be beyond the "native capacity" of such workers.[29]

Thus, Herrnstein's analysis of increasing economic inequality was informed by two assumptions, one leading to an explanation and the other to a justification. First, that the most intelligent people would unquestionably gravitate toward the highest paying professions, whatever they might be. And second, that such a steep gradient of gain would channel the most capable people into those roles most beneficial to the society. Dramatic levels of inequality were, in his analysis, both inevitable and desirable.

There is ample reason to believe that the potential of monetary reward has indeed exercised an effect on career direction, especially for well-educated persons from highly competitive schools with multiple career choices at their disposal—those high SAT scorers epitomizing the group that *The Bell Curve* referred to as the "cognitive elite." As the US economy has experienced a generation-long transition from its traditional emphasis on manufacturing, agriculture, and wholesale and retail trade more toward financialization, graduates from the Ivy League and other elite institutions who had once pursued science, medicine, journalism, public service, and education have turned increasingly to the more lucrative opportunities associated with those areas subsumed under what one observer calls the "casino economy": banking, securities, investments, and trading. In 2006 a *New York Times* article, appropriately titled "Lure of Great Wealth Affects Career Choices," reported on the trend of professionals from other fields to "migrate" to Wall Street: newly minted Ph.D.'s, who once would have pursued a career in teaching and research; law school graduates no longer interested in public interest law or government jobs; and graduates from medical school, some of whom "go directly to Wall Street or into healthcare management without ever practicing medicine." In a particularly dramatic example, the article cited a doctor who, two decades earlier, had graduated from Harvard College and then Harvard Medical School, intending to become a "physician-scientist" with the goal of finding a cure for cancer and "even dreaming of a Nobel Prize." As a hematology-oncology specialist earning $150,000 in 1996 ($245,000 adjusted for inflation), he turned to a business consulting firm, eventually becoming a managing director of healthcare investment banking for Merrill Lynch with an annual income in seven figures. In the 2011 film *Margin Call*, a taut drama about the actions of a Wall Street investment bank during a 24-hour period after one of its junior analysts anticipates the imminent financial collapse, a middle-of-the-night meeting to decide how to react provides a realistic reminder of the finance industry's attraction for the intellectual elite, no matter their original field of study. Asked by his superiors to describe his background, the analyst responds that "I hold a doctorate in engineering, specialist in propulsion, from MIT" and then, prompted to elaborate, explains that his "thesis was a study in the way that friction ratios affect steering outcomes in aeronautical use under reduced gravity loads." "So you are a rocket scientist?" asks the impressed manager running the meeting. "I was ... yes," is the reply; "it's all just numbers really, ... and ...the money is considerably more attractive here." According to Markovitz, "entire groups at major banks" have become "dominated by physicists, applied mathematicians, and engineers, many with Ph.Ds."[30]

Indeed, "considerably more attractive" is a gross understatement; in the last couple of decades the financial industry has become the source of previously unimaginable fortunes unrelated to the progress of the economy as a whole. Two years before the Great Recession, James Simons, a hedge fund manager—instructively, the chair of the math department at Stony Brook with an undergraduate degree in mathematics from MIT and a Ph.D. from Berkeley before becoming a "financial engineer"—took home 1.7 billion dollars; it would take more than 26 years for someone making the median personal income in the United States at the time to earn what Simons received every hour. A year later, after multi-billion dollar bailouts of the financial companies and the loss of millions of jobs, Simons's income grew to 2.5 billion dollars; now his hourly pay of 1.2 million dollars was equal to what someone with the median personal income would take more than 40 years to earn. Even this staggering sum paled in comparison with the amount made that year by John Paulson, who banked 3.7 billion dollars by short-selling the subprime market, while 2.2 million households were faced with foreclosure. By 2018 the top 25 hedge fund managers were making an *average* of 850 million dollars. Though not as handsomely compensated as hedge fund managers, other finance professionals—bankers and traders—have received, in addition to their substantial salaries, regular annual bonuses of millions of dollars both before and after the recession in which their own companies lost billions. In 2014 a young derivatives trader, no longer comfortable with what he called a "wealth addiction," estimated that "90% of Wall Street feels like they're underpaid," describing how his co-workers were "pissed off" at their 2 million dollar bonuses—he himself had once been furious that his own bonus was only 3.6 million—because their bosses were getting 150 million.[31]

Such lucrative possibilities have exerted an understandable effect on the choices made by the cognitive elite, the people expected to become the meritocrats in meritocracy. In 2014, 31 percent of Harvard graduating seniors went on to positions in finance or consulting; a year later more than a third did so. Every year between 2000 and 2010, at least one third of Princeton graduates who took jobs (as opposed to graduate school) entered the financial services industry with a high of 46 percent; if one adds consulting, the percentage is never less than 60 and often considerably higher. At the University of Pennsylvania, almost half of 2016 seniors chose careers in finance or consulting, with 29 graduates going just to Goldman Sachs and another 26 to JP Morgan Chase.[32] It is difficult to believe that large proportions of the nation's most intelligent 22-year-olds suddenly found an irresistible interest in balance sheets, assets, and liabilities, and the more likely explanation for these statistics is that paychecks have taken precedence over passion. And even these figures do not include the many graduates who go first to law school or MBA programs before winding up eventually in finance and consulting. Herrnstein's Skinnerian assumption is undeniably accurate: in making a career choice, the cognitive elite follow the money.

The second assumption, however—that such dramatic levels of inequality are ultimately beneficial—is more questionable. In a well-organized society, of course, it is not only sensible but inevitable for differential rewards to function as incentives, ensuring that talent is utilized both efficiently and effectively; the prospect of greater compensation, whether in money or status, can ensure that people with the right combination of ability and perseverance take on difficult tasks that need doing. But the incredible fortunes now available to the highest earning professions have too often resulted not just in little social benefit but significant harm to the society and the economy.

There is even an argument that the prospect of such large incomes has been detrimental to some of the meritocrats who enjoy them, dissuading them from pursuit of their authentic interests. Some people regard their work as a "vocation," from the Latin verb "vocare," meaning to call or summon; they may feel "called" to preaching, writing, teaching, painting, building, and other activities, all pursued not out of financial interest but for intrinsic reward. As Markovitz's analysis observes, "Work pursued authentically, as a vocation that reflects the worker's true interests and ambitions can be a site of self-expression and self-actualization," one that "integrates work with the other parts of a person's life into an integrated whole." The great dancer Martha Graham, for example, was once asked why she chose to be a dancer and famously responded, "I did not choose. I was chosen." And an in-depth study of eminent scientists described the "driving absorption" that led them to work "long hours for many years, frequently with no vacations … because they would rather be doing their work than anything else." But, Markovitz notes, the demands of meritocratic success trap superordinate workers, precluding them from pursuing work as a vocation and tending "inexorably toward alienated self-exploitation"; the same capitalist affliction "that Marx diagnosed in exploited proletarian labor in the nineteenth century" has been shifted "up the class structure." Although he later bought into Herrnstein's Skinnerian argument, Charles Murray himself once seemed to appreciate the importance of following a calling: a few years before publication of *The Bell Curve* he opposed raising teachers' salaries on the grounds that an increase would only attract applicants "who are 'in it for the money'" instead of the "able and dedicated career teachers …[who] could be making more … if they chose" but preferred the "intrinsic" rewards of teaching.[33]

In any event, Herrnstein's argument for the salutary effect of extreme inequality was based solely on the claim that such generous rewards would attract members of the cognitive elite to the roles of greatest benefit to the society, thus ensuring that the best intellects would serve the collective welfare. The question is whether the actual professions now providing such incredible incomes have in fact played that role.

There is widespread agreement on what those professions are. In *Tailspin*, his compelling account of "The People and Forces Behind America's

Fifty-Year Fall," the attorney and journalist-entrepreneur Steven Brill traced the increase in economic inequality specifically to the gigantic incomes in three often overlapping areas: financial engineers and consultants; corporate executives; and corporate lawyers and lobbyists. Jonathan Rothwell, the Principal Economist at Gallup, listed the same groups as those who "have contributed the most people to the 1 percent [of top earners] since 1980."[34]

Finance

Of these three professions, finance has produced both the highest individual incomes, as the hedge fund managers cited earlier demonstrate, as well as the greatest proportion of the superrich; recent studies have found that between a quarter and a fifth of the richest Americans have made their fortunes in finance, "especially hedge funds and private equity," and the sector accounts for 40 percent of Americans with investable assets of more than 30 million dollars. The finance professionals in the United States once performed a useful service, helping to channel capital in productive directions, increasing the efficiency of markets, and enjoying modest rewards commensurate with their value to the society, but their incomes hardly compared to the wealth of tycoons and owners of natural resources; when the famous financier and banker J.P. Morgan died and his estate was revealed, the steel magnate Andrew Carnegie remarked that he hadn't known Morgan wasn't really a rich man. And half a century ago "Wall Street" referred to a number of small private partnerships that specialized in purchasing equity securities from growing companies and immediately selling them to investors, thus putting their own money at risk and often sitting on the boards of companies they underwrote in order to be fully informed about the investment. In 1970, however, beginning with Merrill Lynch, one after another of these firms decided to go public, not only instantly making their partners fabulously wealthy but now allowing them to take risks with shareholders' money rather than their own.[35] The year before *The Bell Curve* was published, former Harvard University President Derek Bok summarized the outcome, writing that

... much of what transpires in Wall Street seems to go beyond socially productive activity and resembles some sort of casino to accommodate clever people searching for short-term gains. Moreover, however useful financial services may be, the sheer number of highly educated professionals engaged in selling bonds, analyzing stocks, talking with clients, and looking for market anomalies to exploit seems well in excess of any contribution they make to the long-term prosperity of the nation.[36]

Indeed, in the years since Bok's observation, members of the cognitive elite have employed their superior intellectual abilities to devise new and creative strategies for enriching themselves while wreaking havoc on the economy in general. Financial analysts and traders have created one instrument after another designed to produce wealth untethered to hard assets such as buildings, factories, or anything of material value—the economic equivalent of abstract art largely responsible for producing the Great Recession. Instead of directing capital toward the production of goods and services, credit default swaps, mortgage-backed securities, and other complex financial derivatives became the product—not the means to an end but the end itself. By enabling creditors to take out insurance policies for more than the amount actually loaned to a troubled company, some of these instruments created a perverse incentive for hedge funds to push these companies into bankruptcy; after all, the insurance payout would provide more profit than repayment of the entire principal. Some hedge fund managers didn't even bother with computer algorithms or other calculations, merely taking money from clients and passing it off to another fund while skimming a substantial advisory fee off the top. Nor did the recession, in which their own firms lost billions of dollars while so many people of lesser means lost their jobs and homes as a result of these exotic instruments, prevent the financial professionals from continuing to enjoy incredible windfalls. In 2009, nine of the largest recipients of federal bailout money—the same banks and Wall Street firms that had enriched their employees while feeding the nation's economy into a meat grinder—shelled out bonuses of more than a million dollars each to 5000 of their analysts and traders. In 2015, just the annual bonus pool for Wall Street employees came to more than double the combined income for the entire year of all workers earning the minimum wage.[37]

In another financial ploy, so-called private equity firms (as opposed to publicly traded, which are subject to greater regulation and reporting requirements) specialized in purchasing companies, burdening them with as much debt as possible—not to grow the business, reinvest in infrastructure or equipment, hire more people, or otherwise improve prospects, but to pay themselves huge managerial fees quickly returning many times their original investment—and then, when the company has been essentially looted of all financial value, declaring bankruptcy, leaving thousands of workers not only jobless but deprived of the earned pension they had once thought secure. Gordon Gekko, the protagonist of Oliver Stone's *Wall Street* who plots the predatory takeovers of otherwise functioning companies so that he can profit by wrecking them, may have been a fictional character, but his financial machinations come right out of the private equity playbook. The bankruptcies of seven major grocery chains since 2015 involving 125,000 workers, and nine of the ten largest retail firms in 2017 all involved private equity firms. In some cases, jobless workers were not the only ones adversely affected. In 2011, Manor Care, for example, a chain of nursing homes was purchased by the massive private equity firm, the Carlyle Group, which promptly extracted 1.3 billion dollars for investors; when the chain was driven into inevitable bankruptcy a few years later, it was so understaffed that some patients wallowed in their own filth, and rooms were overrun with roaches and ants. Even the success stories—the businesses putatively "rescued" by private equity—combine outrageous profit for the investors with massive loss of jobs for the employees. When Hostess Brands, for example—the manufacturer of Twinkies and other iconic snack cakes—was struggling to survive, the company was bought for 186 million dollars by two private equity firms, which quickly arranged for it to borrow more than a billion dollars for distribution to the investors; Leon Black, co-founder of one of the two firms, received 181 million dollars, while one of Hostess's plants was shuttered and the number of people employed by the company went from 8000 to 1200.[38] As with the bankers and traders, the cognitive elite who dreamed up these schemes prospered while so many others were left in financial ruin.

Corporate Management

Corporate executives constitute another group clustered at the very top of the income distribution. According to a labor journalist specializing in inequality, the "annual jackpots" enjoyed by CEOs comprise "the single largest contributor to the skyrocketing income of America's top 0.1 percent" since 1979, representing 44 percent of their growth; together with the top financial professionals, the two groups account for two-thirds of the increase.[39] In the United States such businessmen (and more recently women) have traditionally enjoyed heroic status. Notwithstanding Marxist theory, most Americans believe that business tycoons represent the true force for historical change—and often for the better, producing the consumer goods and services that improve quality of life for everyone while creating the gainful employment for masses of people that enable them to purchase what they have produced. As "captains of industry," corporate executives are the closest civilian equivalent to military officers. Though working in the private sector, they lead organizations that constitute what Franklin D. Roosevelt referred to as "a public trust," serving not just shareholders but employees, customers, and communities.

Unlike finance professionals, who have only recently ascended to the highest income levels, corporate executives have historically been the most highly paid people who worked for a living. Indeed, social scientists typically considered them well deserving of such generous remuneration, given their contributions. At the dawn of the twentieth century, William Graham Sumner, the nation's first ever professor of sociology, explained that millionaires—all heads of business at the time—were "a product of natural selection, acting ... to pick out those who can meet the requirement of certain work to be done .... They get high wages and live in luxury, but the bargain is a good one for society." In 1940, E.L. Thorndike, one of the first psychologists to be elected to the National Academy of Sciences, argued that superior intellects should receive whatever they wanted, citing in particular executives of large businesses, who, he claimed, merited much greater salaries than they were receiving, considering the "value of the services" they provided. "Whatever will put great managerial ability at work should be offered," Thorndike concluded—"money, power, prestige or whatever else is required." And C. Wright Mills's mid-1950s classic, sociological study of influence in mid-twentieth-century America, *The Power Elite*, described a common opinion of the richest corporate executives at the time as "responsible trustees, impartial umpires, and expert brokers for a plurality of economic interests, including those of all the millions of small property holders who hold stock in the great American enterprises, but also the wage workers and the consumers who benefit from the great flow of goods and services."[40]

Although corporate executives have thus long enjoyed some of the highest earnings, their present compensation has nevertheless grown exponentially in comparison with the past. Mills's study noted that the highest paid executive in 1952—indeed, the highest paid individual at a time when, as Mills observed, "All the men ... of great wealth are now identified with large corporations"—was Charles Wilson, the CEO of General Motors, who made $581,000 in salary and bonuses or about $5.5 million adjusted for inflation; average CEO salary at the time was $100,000 or about $950,000 in contemporary dollars. By 2017 the highest paid executives took home more than $100 million apiece, and the average pay for CEOs of the largest 350 companies was 18.7 million.[41]

Nor is it the case that these increases have merely reflected a broader trend of rising salaries for all a corporation's employees; according to a study by the Economic Policy Institute, between 1978 and 2017 CEO pay rose by 1000 percent while the average employee salary increased 11 percent. In the 1950s and early 1960s, a CEO typically earned about 20 times the salary of the firm's average employee, a ratio that had long been considered the optimal maximum. In an earlier era, J. P. Morgan had famously insisted on never paying an executive more than 20 times the earnings of a company's lowest paid employee. And in 1984 Peter Drucker, widely acknowledged as the founder of modern management, called for a "voluntary limitation" on executive pay of at most 20, and perhaps even 15, times the pay of the rank and file; years later he was quoted in a letter submitted to the SEC as having "advised managers that a 20-to-one salary ratio is the limit beyond which they cannot go if they don't want resentment and falling morale" in their company. By 2017, however, the ratio of CEO compensation to the average worker's salary had grown to 361—every single day the former's income amounted to approximately what the latter received for the entire year—but typically much higher for larger corporations; the 102 million dollar pay package for the CEO of First Data, for example, was more than 2000 times the median compensation of all other employees.[42] But even these dramatic statistics do not indicate the full extent to which executive compensation has increased, since the marginal tax rate for the highest incomes in 1952 was more than 90%—for every dollar of salary over $100,000, a taxpayer got to keep less than 10 cents—whereas the highest marginal tax rate is now only 37%. Presuming no extraordinary deductions, in 1952 after taxes Charles Wilson would have been able to keep around $115,000 of his $581,000 income. Nor did the earlier rate produce the howls of outrage later elicited by anything greater than the current one; in the post-war world an almost confiscatory marginal tax rate on the highest incomes was accepted as a social responsibility without protest by the people fortunate enough to make that much money.

In addition to making considerably more money, corporate executives now tend to arrive at their position through a much different career path than in the past. It was once not uncommon for managers from many different backgrounds to work their way up the corporate ladder, spending much of their working life with the same organization; Wilson, for example, joined a General Motors subsidiary as an engineer and sales manager in 1919, not rising to president of the company until 22 years later. Having come up through the ranks, such executives tended to rely for advice on the many people in the middle management positions they had once occupied. In contrast, most contemporary chief executives typically come from an elite institution; according to Markovitz, half of America's corporate leaders attended one of twelve universities and typically hold an MBA or similar postgraduate degree. When one of these corporate heads needs assistance, they look to the professional management consultants from identical backgrounds, the ranks of middle management having been hollowed out and their salaries essentially transferred upward.[43]

While Herrnstein's second assumption is obviously risible with respect to financial professionals—providing obscene incomes for these members of the cognitive elite has hardly ensured the use of their abilities for the greater social good—the case of business executives is more complex. Certainly, there are well run corporations that compensate their executives in a manner reflecting the company's contributions to consumers, employees, shareholders, and the larger economy. However, the evidence strongly suggests that such ideal results are as much the exception as the rule.

A number of studies over the past three decades have found little rational justification in most cases for the outlandish salaries enjoyed by corporate executives. In 1991, when the ratio of CEO pay to that of the median employee was less than half of what it would later become, Graef S. Crystal a prominent "executive compensation consultant"—i.e., a professional hired by large companies to assist in the design of appropriate compensation packages for top level management—authored a book whose subtitle crisply summarized its conclusion: *In Search of Excess: The Overcompensation of American Executives*. In blunt prose Crystal outlined the problem of "bloated pay packages," resulting in "huge and surging pay for good performance, and huge and surging pay for bad performance, too." Derek Bok's 1994 study of highly paid professions acknowledged that "in earlier generations" executives had earned much more than others, when "there was at least an observable link between reward and talent, entrepreneurial success and social welfare." But, he wrote, "scholarly analysis" now indicated that "performance pay for top executives has turned out to be a sham and an embarrassment," so that "the compensation actually paid to CEOs bears very little relation to the record of their companies" in terms of corporate value. Moreover, he found no relation to larger civic goals—no link between "the pay executives receive" and "the contribution they make to social welfare or economic growth."[44]

A decade later Lucian Bebchuk and Jesse Fried, Harvard law professors specializing in business and the economy, published another instructively titled book: *Pay Without Performance: The Unfulfilled Promise of Executive Compensation*. While both this book and the earlier volume by Crystal detailed the numerous devices employed to further enrich executives beyond their huge base salaries—guaranteed bonuses, stock options rigged to pay off even when the stock price plummets, golden parachutes ensuring windfall payments for departure even when caused by abysmal performance—Bebchuk and Fried went on to explain how compensation packages were not only unrelated to performance but also deliberately structured to "camouflage" the true amount paid, both during and especially after the executive's employment. Defined benefit retirement pensions, guaranteeing 7-figure annual incomes—unlike the defined contribution plans typically offered to other employees, which are exposed to the risks of investment—can wind up providing greater income over time than the executive had earned during his or her actual employment; lucrative consulting contracts—not for work actually performed but for an "availability" to consult that is rarely invoked since new CEOs are typically not inclined to seek advice from their predecessors—add more substantial sums; perks, such as access to corporate aircraft, chauffeured cars, paid assistants, apartments, and more are also worth huge amounts of money. According to Bebchuk and Fried, unlike the more traditional methods, none of these forms of "stealth compensation" must be disclosed through the usual channels, thus keeping them hidden from the public in general and shareholders in particular.[45]

All three of these books culminated in recommendations for reforming the process of determining executive pay. Bok, for example, proposed a number of changes designed "to tie compensation more closely to performance." Though cognizant that no remedy could "cure all the ailments," he was certain that the "worst abuses" could be curbed—"fewer CEOs taking home millions of dollars as their companies wallow in substandard performance."[46]

Such expectations have proved naïve. While there is little data on the systematic relationship between CEO compensation and corporate performance, there has been no shortage of instances that Bebchuk and Fried described as "generous treatment even in cases of spectacular failure"—companies that went down the financial tubes, costing shareholders their investment, and sometimes employees their jobs, while the CEOs who presided over these fiascos received princely sums quite apart from the "camouflaged" forms of compensation, not in spite of their failure but because of it—the necessary price for getting rid of them. Shortly after their book appeared, for example, Alan Mulally, Ford Motor

Company's newly hired CEO, received 28 million dollars for his four months in the position, after the company announced a 12.7 billion dollar loss and plans to close plants and cut more than 30,000 hourly workers. Even that fortune paled in comparison with two other hastily arranged exit packages around the same time: Robert Nardelli, the CEO of Home Depot, received 210 million dollars after the company's stock dropped and it lost market share to Lowe's, and Henry A. McKinnell left Pfizer 200 million dollars richer after its share price fell 40% and thousands of employees were terminated. Though not quite as magnanimous, Hewlett Packard could claim the unique distinction of paying a succession of CEOs to retire: after a dismal record Carly Fiorina was paid 21 million dollars to leave and was replaced by Mark Hurd, who received 12.2 million dollars to resign in the wake of accusations of sexual harassment and financial improprieties, and was replaced by Leo Apotheker, who was given a 13.2 million dollar exit package less than ten months later, after the company lost more than 30 billion dollars in market capitalization under his leadership. And in the most recent example, after Boeing's 737 MAX disaster—two crashes that cost 346 lives and erased 11 billion dollars of market value—Dennis A. Muilenburg was ousted as CEO, leaving with more than 62 million dollars in stock and pension awards.[47]

Nor do any of these examples, and numerous similar ones, include the many cases in which CEOs created their lucrative retirement package through ethically questionable methods. Bebchuk and Fried noted that the executives of the 25 largest firms to go bankrupt in the first few years of the twenty-first century sold almost 3 billion dollars of stock shortly before their companies tanked. Hundreds of other executives at firms whose share price plummeted by 75 percent or more conveniently unloaded 23 billion dollars' worth of stock altogether just before the descent began. In some of these cases, while cashing out their own holding, the firm's management prohibited employees from selling the stock out of their 401(k) accounts.[48] And of course the CEOs of financial firms earned obscene amounts of money by selling products they knew to be defective, shattering the global banking system, and requiring the government to save them from closing their doors—a taxpayer expenditure that did not deter them from awarding themselves 7-figure bonuses at the same time.

Although the coronavirus forced many large businesses, especially in retail, to declare bankruptcy, it did not inhibit executives from arranging substantial bonuses for themselves shortly before closing their doors for good. In the first six months of the pandemic, 18 large companies collectively distributed 135 million dollars to their executives just before requesting court protection from their creditors. Five days before filing for Chapter 11, J.C. Penny, for example, awarded 7.5 million dollars to its four top executives while closing approximately 150 stores and eliminating thousands of jobs.[49]

While individual exceptions doubtlessly exist, just as with the Wall Street bankers and traders, for high-ranking executives it is again difficult to find support for Herrnstein's contention that such a steep gradient of gain, producing previously undreamed-of incomes for select members of the cognitive elite, has resulted in considerable benefit for the society overall.

Law

A final profession disproportionately represented among the highest incomes is corporate lawyers, both those who work as lobbyists as well as in-house counsel. The presence of attorneys in the ranks of the super-rich represents a fairly recent development. Not that long ago, before the legal profession became the butt of jokes based on their ethical shortcomings, lawyers, like financiers, were comfortably but not excessively compensated in accord with their service both to individuals and the society. But quite apart from how much they earned, there was a sense that lawyers played a crucial role in ensuring stability and fair play—perhaps an avenue to material prosperity but also the bedrock of the society's moral order. As C. Wright Mills observed in mid-century, that role began to change with the "incorporation of the economy," as more lawyers, especially in urban areas, "made business counseling the focus of their work, at the expense of traditional advocacy." In particular, corporations turned to lawyers for assistance in "minimizing ... [their] tax burden, ... controlling government regulatory bodies, [and] influencing state and national legislatures."[50]

Nevertheless, until the 1970s, according to Steven Brill—the attorney who founded both *The American Lawyer*, a magazine covering the business of law firms and lawyers, and the cable channel, Court TV—law remained "a relatively sleepy profession"; even at white-shoe firms new hires began with a salary only 20–25 percent higher than the average income. Starting in the 1970s, however, the "demand for lawyers exploded" in response to corporate interest in mergers, tax shelters, and a host of new government regulations involving consumer products, discrimination, worker safety, and protection of the environment. As a result, Brill noted, the new lawyers were concentrated in "firms that served large corporations and were prepared to pay skyrocketing salaries to attract the best talent"—the strongest students from the most competitive law schools—and by 2016 the partners at the top corporate law firm earned an average of 6.6 million dollars. As Herrnstein had predicted, the cognitive elite followed the money: Markovitz reports that more than half the partners at the five most profitable firms graduated from one of the top ten law schools; for the single most profitable firm that figure rises to 96 percent. By 2015, almost 60 percent of the 2010 graduates from Yale Law, a school previously known for graduates who entered public service, were employed by business firms; for 2015 graduates from Columbia Law, a year later more than 70 percent had taken such positions.[51]

The purpose of this cadre of lawyers was to prevent government from functioning in any way that might be detrimental to corporate interests. Brill described, for example, how, after first being proposed in the Federal Register, Occupational Safety and Health regulations concerning hazardous chemicals in the workplace were delayed for years by lawyers who offered thousands of pages of comment—each of which needed to be read and considered—scheduled countless meetings, appealed rules in court, quibbled over the meaning of common words, and generally did everything possible to throw a monkey wrench into the process. Thus, shortly after OSHA began operation in the early 1970s it took about a year to complete the review process resulting in a 10-page-long rule for a particular chemical, while in 2016 a rule about a different chemical took 19 years to write and consumed 604 pages; as part of the process OSHA estimated that the latter rule would have prevented more than 579 deaths a year. The corporate lawyers "swarming the process," as Brill put it, have had the desired effect: even though many hundreds of new chemicals are introduced into the workplace each year, since its inception OSHA has been able to issue regulations in but thirty cases, only three of them since 1997.[52]

In theory, of course, the interests of consumers, employees, and the public are supposed to be represented by lawyers for either the government or, less often, a non-profit organization; the assumption underlying the legal system is that the "correct" result is most likely to occur when advocates for each side enjoy an equal opportunity to make the best case possible. In practice, however, lawyers in the public sector are almost always outgunned in both quantity and quality by those representing corporate interests. In his 1994 study of highly paid professions, Derek Bok provided a concise description of the problem:

> The ablest lawyers usually go to established firms where they frequently litigate and negotiate with a much less experienced government attorney or with a solo practitioner representing a private claimant. ... In all these circumstances, so long as the most promising young lawyers choose overwhelmingly to serve large corporations, continuing to add more and more exceptional talent to the profession may help to make legal encounters more unequal and to increase the odds of prevailing for reasons other than the true merits of one's case. If so, the influx of exceptional talent may succeed not in furthering justice but in magnifying the human imperfections of our legal system so as to diminish, rather than enhance, the welfare of society.[53]

Since Bok's warning, the malign influence exerted on public policy by corporate lawyers from elite schools has only increased, especially with the exponential growth in their deployment as Washington lobbyists from a cottage industry to an immense enterprise, turning "K Street" into a metonym for the entire operation. In 1971, 175 firms employed registered lobbyists in the capital; by 2016 more than 7700 corporations and trade associations did so at a cost of more than three billion dollars. The overwhelming number of these lobbyists work for business rather than for public interest groups; none work for unions. The two most powerful lobbies represent health care and finance, the former industry employing five lobbyists for every member of congress, the latter spending a million dollars per member. From the corporations' perspective, this was money well spent; whether the issue was financial reform, the tax code, consumer rights, or labor relations, it was much cheaper to pay huge amounts of money to an army of top-notch lawyers and lobbyists to chip away at the substance of proposed legislation—inserting new provisions, adding qualifiers, injecting exceptions, tweaking definitions, postponing starting dates, introducing vague language ("reasonable" interest rates) that can be subsequently litigated—rather than obey it. The "Banking Act of 1933," for example, also known as "Glass-Stegall"—the financial bill passed just after onset of the Great Depression, establishing the Federal Deposit Insurance Corporation among other reforms—was all of 33 pages long; thanks to the efforts of 2000 lobbyists—four for every member of both houses of congress—"Dodd-Frank," the analogous attempt at modest reform of banking in the wake of the 2008 Great Recession (shorthand for the "Dodd-Frank Wall Street Reform and Consumer Protection Act") consumed 2319 pages with unnecessary complications and impenetrable prose, not to mention hundreds of provisions dependent on the lengthy and complicated rule writing process that would take years to complete. "We have three lawyers total" working on the bill, commented the legislative director for a large consortium of non-profit groups; the banks have three lobbyists "working on a paragraph."[54] In another example of complexity introduced by lawyers/lobbyists mainly for corporate benefit, the original income tax code, passed in 1913, was 27 pages long; by 2017 it had grown to 6550.

Well compensated lobbyists miss few opportunities to pursue corporate interests even at a cost to public welfare. For most of the nation the first coronavirus stimulus bill was an emergency measure designed to alleviate some of the worst effects of the pandemic on the economy. Instead of a response to tragedy, however, lobbyists saw the bill as an opportunity—what the *New York Times* called an "irresistible target"—leading to "a frenzied effort to insert into the must-pass legislation provisions their clients wanted," many of which "were largely unconnected to the coronavirus crisis." As one congressional reporter put it, "lobbying firms of all stripes lined up at the trough": undeterred by social distancing measures, they pressed their cause by phone and email, achieving tweaks to the tax code for wealthy investors in real estate and energy, banks, and large hotel chains, which these interests had sought long before anyone had ever heard the word "covid." According to a law professor at the University of California, Irvine specializing in taxation, many of the tax benefits in the stimulus bill were "just shoveling money to rich people."[55]

As the stimulus bill indicates, together with bankers and accountants, many elite lawyers are engaged in what Markovitz calls the "income defense industry," enriching themselves by protecting "still richer people's fortunes against government encroachment," thereby thwarting any attempt by the state to regulate great wealth. "The trust and estates bar alone," he points out, comprises over fifteen thousand lawyers, "whose work no doubt justifies the observation by Gary Cohn, chief economic advisor to President Trump during the first year of his administration, that "only morons pay the estate tax." And specialists in tax havens have allowed those with more than thirty million dollars of investable assets to move collectively some eighteen trillion dollars' worth offshore. As a result of these and other maneuvers, at the same time that the share of national income enjoyed by the richest individuals doubled, their tax rate fell dramatically. According to ProPublica, a non-profit organization of investigative journalists, in some years a number of the wealthiest people in the United States paid no income tax at all, and between 2014 and 2018 the 25 richest averaged a tax rate of 3.7 percent.[56]

There is even reason to believe that the exorbitant earnings enjoyed by corporate lawyers played a significant role in the justice department's failure to pursue criminal charges against the finance executives responsible for the 2008 economic debacle despite ample evidence of their fraudulent behavior. Such reluctance to prosecute corporate misbehavior is relatively recent. Three decades ago hundreds of people associated with, first the savings and loan crisis, and then the "junk-bond" market were prosecuted and received stiff fines and jail sentences. Then, when the tech bubble burst at the beginning of the present century, government attorneys did not hesitate to seek harsh penalties against superrich executives who had looted their companies while shareholders lost their investment and employees their jobs. Charged with felonies such as bank fraud, securities fraud, insider trading, and perjury, a number of CEOs received not only substantial fines, which they had to pay personally, but lengthy prison terms: John Rigas (Adelphia Communications), who was sentenced to 15 years; Joseph Nacchio (Qwest Communications) to 6 years; Dennis Koslowski (Tyco International) to 8–25 years; Samuel Waksal (Imclone Systems) to 7 years; and two executives at Enron—Kenneth Lay, who died three months before his scheduled sentencing, and Jeffrey Skilling, whose original sentence of 24 years was reduced to 14 on appeal. In contrast, after the Great Recession 49 financial institutions paid a total of 190 billion dollars in fines for various misdeeds, but no individuals were charged; crimes had been committed, but there were no criminals. And since the money was paid by the corporations rather than the executives, making a substantial portion tax deductible, it actually came from the pockets of shareholders and taxpayers.[57]

What accounted for this dramatic change in response to corporate misbehavior? The Pulitzer Prize winning business journalist, Jesse Eisinger, has argued that young justice department lawyers have become reluctant to pursue corporate officials too vigorously, even when there seemed to be a prima facie case of criminally fraudulent behavior, lest such prosecution impedes their own prospects once their stint in the public sector has ended; knowing that they would eventually leave public service, they wished to remain hirable at high paying firms. According to Eisinger, a newly appointed assistant prosecutor is typically the product of an "elite" institution, an ambitious student who has spent "endless hours slaving to achieve the highest grades." But having landed a prestigious position at the Department of Justice,

> at that point, then, their formula for success has altered. They are not trying to please the powerful. If they do their job, they will displease them. To prosecute those sitting in corporate boardrooms, the young government litigators must become class traitors, investigating and indicting people very much like their mentors, peers, and friends. ... A corrupt politician excites in upstanding prosecutors a sense of outrage.

By contrast, a well-mannered and highly educated executive seems like someone who wouldn't knowingly do something wrong.

Until fairly recently, top level corporate law firms did not represent their clients in criminal matters, but now at settlement negotiations prosecutors, whose salaries were set at a maximum of $160,000, were sitting across the table from professional counterparts making 20 times that amount. As Eisinger explained, in these circumstances prosecutors want to "appear tough" to the defense lawyers, to "dazzle them with their knowledge of legal precedent, mastery of details and bargaining skills. But young prosecutors also want their adversaries to imagine them as future partners. They want to be seen as formidable but not unreasonable. They want to demonstrate that they are people of proportion," and by doing so, they "set themselves up for lucrative careers in the private sector."[58] Indeed, Eisinger named one prosecutor after another who negotiated a deal for high-profile white-collar crime and then accepted lavishly paid partnerships at prominent white-shoe firms, while the few government lawyers who had taken a more aggressive approach found themselves blackballed. Out of probable self-interest it seemed that one sector of the cognitive elite decided to give another sector a pass on criminal behavior, choosing to protect the interests of the wealthy rather than uphold the society's moral order.

Markovitz's conclusion about the contributions of all these extraordinarily well-paid "superordinate" workers did not mince words. "The elite's true product," he wrote,

> may be near zero. For all its innovations, modern finance seems not to have reduced the total transaction costs of financial intermediation or to have reduced the share of fundamental economic risk borne by the median household, for example. And modern management seems not to have improved the overall performance of American firms (although it may have increased returns to investors). More generally, rising meritocratic inequality has not been accompanied by accelerating economic growth or increasing productivity.[59]

Thus, the evidence strongly suggests that Herrnstein's second assumption is not merely unfounded but diametrically opposed to what has actually occurred. While there have undoubtedly been some exceptions, the availability of massive financial rewards has not ensured that superior intellect is systematically directed toward socially beneficial activity. Instead, it has often encouraged members of the cognitive elite to engage in what is essentially insider looting, allowing them to construct what Dennis Kelleher, the President of Better Markets—an organization founded in the wake of the Great Recession to promote financial reform—has aptly described as a "wealth extraction mechanism for the few rather than a wealth creation system for the many."[60] Highly intelligent people have raked in obscene amounts of money for behavior that is at best socially unproductive and at worst highly detrimental, while producing the largest redistribution of wealth upward in human history. Not only is there no moral justification for such extreme inequality, it is not possible to rationalize the second Gilded Age on the grounds that the cognitive elite have received remuneration appropriate to their contributions.

Naturally the fact that such incredible private rewards bear no, or even a negative, relation to public welfare is not to argue that differences in income serve no useful purpose. Scarce natural talent certainly merits greater compensation when directed to those activities that best serve the common good. However, at present the relationship between monetary reward and social contribution has fractured; not only do the careers most responsible for extreme inequality often fail to serve the common good, but to the extent that the prospect of riches does in fact attract the most intellectually capable, it only dissuades such talented persons from pursuing professions with the greatest potential of achieving that laudable goal.

Innate Inequality: Intelligence and Human Value

Even if "it is granted that differentials in economic rewards are morally justified and socially useful," wrote the ethicist and theologian, Reinhold Niebuhr, in his influential 1932 book, *Moral Man and Immoral Society*, "it is impossible to justify the degree of inequality which complex societies inevitably create.... The literature of all ages is filled with rational and moral justifications of these inequalities, but most of them are specious ... clearly afterthoughts." In a meritocracy the rich must appear worthy of their good fortune in order to be perceived as in some sense legitimate; if the financial rewards enjoyed by so many of the cognitive elite are far from commensurate with the social value of their accomplishments, then a different justification must be found. The meme persists that high IQs are a marker of superiority, entitling the cognitive elite to much more of the society's material resources than the lesser endowed, not so much because of what they have done but as a confirmation of who they are: the smartest of the smart, educated at the most prestigious universities, whose position in the top sliver of the intelligence distribution entitles them to a corresponding position in the income distribution. Indeed, in Herrnstein's original article on the subject, he instructed those with a high IQ who wanted to become rich to "not waste your time with formal education beyond high school"; a degree from an elite institution, in this view, might have been merely the suggested attire for a party that the highly intelligent were destined to attend no matter their wardrobe.[61]

Alex Rubalcava, a Harvard student who would eventually go on to a successful career in venture capital, made the case even more bluntly, arguing that the actual substantive tasks performed by management consulting firms and Wall Street investment banks—facilitating mergers and acquisitions, selling shares to the public, and generally helping "wealthy individuals stay wealthy"—could and should be assigned to junior officers or even outsourced, so that these companies could focus all their efforts on what was their true "core competency": recruiting Harvard students. "Remember," he explained, "that companies that do nothing of value must obscure that fact by hiring the best people to appear dynamic and innovative while doing such meaningless work."[62] The quality of financial advice was much less important than the presumed intelligence of the people who offered it.

A number of ethnographic studies have provided ample evidence that Rubalcava's view is widespread. In *Liquidated: An Ethnography of*

Wall Street the anthropologist Karen Ho found—based on both interviews and her own experience as a Princeton student recruited to be a consultant at Bankers Trust—that the major firms directed "Herculean recruiting efforts" toward students from highly selective universities with a particular focus on Princeton and Harvard. At such elite institutions financial firms tended to dominate campus life through their constant presence at career forums, panel discussions, "meet and greets" offering free drinks and hors d'oeuvres, and their "goodie bags" filled with so much paraphernalia that "thousands of students become walking advertisements as their logos disperse into campus life." After joining an investment firm, graduates from these universities receive special training programs that fast-track them for prestigious front office positions, while their co-trainees from "second tier" schools such as Rutgers or NYU are placed in parallel classes slating them for "less prestigious and (much) less well-paid divisions." The fact that the objects of all this attention may have neither technical skill nor business savvy is insignificant; their intellectual pedigree and the superior intelligence it putatively represented is what counts. Top banks and investment firms claim that they have created "the most elite work-society ever to be assembled on the globe," staffed by "the greatest minds of the century," "the smartest people in the world"; they brag that "we hire only superstars," graduates "only ... from five different schools," who constitute "the cream of the crop."[63] Again, it was their intelligence, as indicated by the elite institution which these employees had attended, that proved the worth of their advice and justified their entitlement to huge incomes.

Research by the sociologist Lauren Rivera found a similar dynamic. After graduating from Yale, Rivera spent two years in management consulting before deciding to become a sociologist studying the environment in which she had worked. Using the business connections she had established, Rivera secured a position as an unpaid "'recruiting intern' to help plan and execute recruitment events" for three types of "Elite Professional Service firms"—investment banks, "top-tier law firms" and management consulting firms—in exchange for which she received permission to observe the process and interview the recruiters. The most important single factor for these recruiters was the "prestige" of an applicant's institution, with four universities—Harvard, Yale, Princeton, and

Stanford—enjoying what she called the "*super*-elite" status that placed their graduates at the head of the line; other schools nationally ranked among the top 25 might be recognized as highly selective but nevertheless lacked the door-opening prefix. Again, students at the top schools were courted lavishly; one firm allotted a budget of close to a million dollars per year for recruiting events at one of the super-elite campuses. Neither what applicants studied nor how well they did mattered as much as what school they attended, as employers essentially "outsourced" their screening to admissions committees at elite universities. As Rivera put it, the credential most highly valued by these ESPs "was not the education received at a top school but rather a letter of acceptance from one." Those selected at the end of the process not only enjoyed "unparalleled economic rewards for young employees" with no experience but also received "signing bonuses … as well as relocation expenses." Rivera's investigation, just like Ho's study, found that elite professional firms justified the huge expenses devoted to hiring graduates from super-elite institutions by marketing their employees as the "best and brightest," "likely to become superstars."[64]

Even the outrageous bonuses paid to Wall Street Executives *after* they had presided over their company's financial meltdown were justified by their supposed brilliance. In *Bailout: An Inside Account of How Washington Abandoned Main Street While Rescuing Wall Street* Neil Barofsky, the former federal prosecutor named Special Treasury Department Inspector General to oversee the Troubled Assets Relief Program, described how the insurance giant AIG received 170 billion dollars from taxpayers to avoid collapse and then distributed 168 million dollars in "retention bonuses" to members of its Financial Products Division, "the very unit whose reckless bets had brought down the company." Barofsky was surprised to find that Treasury officials "didn't seem to begrudge the AIG executives the bonuses at all, viewing the payouts as "necessary to keep the 'uniquely' qualified" individuals in position to undo the mess they had created: "The Wall Street fiction that certain financial executives were preternaturally gifted supermen who deserved every penny of their staggering paychecks and bonuses was firmly ingrained in Treasury's psyche." Even after the financial crisis revealed their incompetence, the belief endured that a Wall Street Executive receiving a "$6.4 million 'retention' bonus ... must be worth it."[65]

But bloated incomes are merely the extrinsic manifestation of the cognitive elite's intrinsic value, which, for many scientists, enamored of the importance of intelligence, has always extended beyond the monetary sense. From its inception the IQ test has been regarded by many of its most ardent advocates not as merely a measure of a highly specific sort of cognitive ability—usually defined as involving conceptualization and abstract reasoning—but as an indication of inherent worth. Not only do the cognitive elite deserve more, but their greater intelligence makes them innately more important people, whose lives and wishes matter more than those of the less cognitively gifted. As early as 1920, H.H. Goddard, who had translated the original IQ test—the Binet—from French into English, recommended that "men should be paid first according to their intelligence; and second according to their labor," even for persons performing the same job. While it might seem odd for intelligence to take precedence over productivity, especially in a book claiming to explore the relationship between intellectual ability and "human efficiency," Goddard pointed to the more refined sensibilities of those with greater intelligence as the justification for their material entitlement. Addressing those he considered his intellectual equals, Goddard ridiculed the possibility that someone with less intelligence "could live in your house with its artistic decorations and its fine pictures and appreciate and enjoy those things"; it was, he insisted, "a serious fallacy" to "argue that because we enjoy such things, everybody else could enjoy them and ought to have them." In a slightly less condescending justification, two decades later E.L. Thorndike, the country's most prominent educational psychologist at the time, proposed a precise mathematical system to determine how much weight should be accorded to the desires of each individual; the desires of an average person would count for 100, those of someone of superior intelligence for 2000. Although Thorndike acknowledged that some "men of genius" had sometimes sought "eccentric, ignoble or ruthless satisfaction," nevertheless he thought it imperative to identify such persons as early as possible and "give them whatever they need." And "what they need," he concluded, "is what they themselves desire." For the intellectually superior, no desire was to go unfulfilled.[66]

According to many prominent social scientists early in the twentieth century, being of greater value also entitled the cognitive elite to greater political influence; soon after creation of the intelligence test Charles Spearman even suggested it be used to select only the "better endowed persons for admission into citizenship." More common, however, were proposals to change the rules for eligibility to vote, typically disenfranchising the less intelligent. Goddard, for example, found it "a self-evident fact that the feeble-minded should not be allowed to take part in civic affairs; should not be allowed to vote"—this at a time when the mass testing of draftees had led him to conclude, "beyond dispute" that half the nation was "little above the moron." After warning of "distinctly inferior" immigrants as well as many Hispanics and Blacks who could never be "intelligent voters or capable citizens," Terman called for "a less naïve definition of the term democracy," one that would "square with the demonstrable facts of biological and psychological science."[67] Another psychologist of the time, George Barton Cutten, who went on to become president of Colgate University, happily anticipated that IQ tests would produce "a caste system as rigid as that of India," depriving at least 25 percent of citizens of the ballot. And William McDougall, occupant of the William James chair of psychology at Harvard and arguably the most well-known academic psychologist in the English-speaking world at the time, declared that the franchise "must be denied to those who are obviously unfit to exercise it," a policy he believed should apply to all democracies but especially in the United States, "made up as it is of so many heterogeneous elements," where he anticipated that between a quarter and a third of the adult population would not be allowed to vote.[68]

While such sentiments are now largely rejected in an era more sensitive to individual rights, exceptions remain. Raymond Cattell, author of an enormous body of research in personality, human intelligence, and multivariate methodology and the seventh most highly cited psychologist of the twentieth century, supported restriction of the franchise throughout his lengthy career. In the 1930s he thought it "goes without saying" that the less intelligent should be prevented from voting and expected no opposition to such a proposal since those affected "seem to realize that their greatest happiness lies in a benevolent dictatorship." Half a century later Cattell was outraged to realize that the latter expectation was clearly no longer tenable and railed at what he called "robbery ... by the ballot box," the use of the franchise by the "less gifted" to usurp the prerogatives of their intellectual superiors. To rectify this injustice he recommended various possibilities: a minimum IQ score, reducing the electorate to 60–75 percent of its present size, or an "explicit weighting of the votes of individuals according to their intelligence, sanity, and education," a proposal he justified by comparing two "personal acquaintances": a famous "classics professor ... with a deep grasp of the political and social wisdom of the ages" and "an ordinary person who did some gardening." The present practice of democracy, Cattell complained, allowed the latter's opinion "to completely cancel" the former's "long sighted contribution to the community"; the society could not survive, he concluded, "if it gives equal voting powers to individuals so disparate." (In all likelihood, the classics professor was Revilo P. Oliver, a friend and colleague at the University of Illinois acknowledged by Cattell in print as an influence on his thinking, and arguably the leading Nazi intellectual in the United States at the time, who looked forward to future recognition of "Adolf Hitler as a semi-divine figure.")[69]

If the more intelligent are of greater value to the society and if, as so many IQ scientists have concluded, intelligence has a substantial genetic component, then it follows naturally that the children of the more intelligent are of greater value than other children. Just as there have been calls to deprive the less intelligent of the franchise, ever since Francis Galton first proposed the concept of eugenics a century and a half ago, there have also been attempts to restrict their reproduction. Galton himself believed that the less intellectually capable would voluntarily accept appropriate limits on their behavior, but those who refused and continued to burden the society with their inferior offspring, he wrote ominously, would be "considered as enemies to the state." And Spearman maintained that test scores should be used to determine "the right of having offspring." Supported by many scientists, one of the major successes of the eugenics movement was the passage of laws authorizing involuntary sterilization of the "feeble-minded," a practice that began early in the twentieth century and continued well into the post-war period.[70]

Similar to restriction of the franchise, involuntary sterilization is no longer acceptable though some scientists have continued to stress the importance of non-coercive measures to stop the supposed dysgenic trend caused by fecundity of the less intelligent. In 1963 the eminent University of Chicago physiological psychologist and pioneer in endocrinology Dwight J. Ingle, a member of the National Academy of Sciences, recommended quarantining those "poorly endowed with intelligence" in specific complexes—low IQ housing—where they would be provided with "an intensive program of birth control." Throughout the next decade he continued to offer various plans for "selective population control," typically by encouraging "barrenness ... among the mentally dull" through subsidized sterilization or unspecified "material rewards." By 1973 he was recommending that a group of professionals—scientists and physicians—determine the "genetic, ... social, economic and behavioral fitness of the individual for parenthood," a procedure that could be enforced by implanting "pellets of antifertility agents under the skin" of a woman who would then "have to apply for a license to have the pellet removed in order to become pregnant."[71] Around the same time the Nobel Laureate physicist-turned-behavior-geneticist William Shockley warned that medical advances were now assuring "to all the privilege of reproducing their kind," leading to proliferation of the less intelligent. To halt this trend he proposed a "Voluntary Sterilization Bonus Plan": in exchange for agreeing to be sterilized a person would receive $1000 for each IQ point below the population average of 100. And because he thought it most important to reach "those who are not bright enough to hear of the bonus on their own," Shockley suggested that "bounty hunters" be paid a portion of the reward for persuading "low IQ high-bonus types to volunteer." Arthur Jensen, too, viewed people with low IQs as "a burden on everyone, a disservice to themselves" and urged that "we should prevent their reproducing." While he offered no specific proposal, he warned of the genetic deterioration from "current welfare policies, unaided by eugenic foresight," clearly implying the necessity for some policy that would limit reproduction of the less intelligent.[72] All three of these scientists emphasized that this dysgenic trend was much more severe within the black community, each claiming that the genetically least capable Blacks were producing the largest number of offspring. Indeed, each of them employed the same Orwellian phrase—"genetic enslavement"—specifically to Blacks, suggesting that their true shackles were now internal and that only control of reproduction could remove them.[73]

Again, it was Raymond Cattell who called for the most extreme measures. For Whites in his scheme, each person was to be assigned "a precise factor of fertility," determining the "desirable number of offspring," and the consequences for those who defied expert advice and brought "mentally backward children into the world in the face of recommendations to the contrary" would include sterilization, payment of a fine and even incarceration. For less capable minorities, however, such individual distinctions were unnecessary. "Where it is obvious that the race concerned cannot hope to catch up in innate capacity," he wrote early in his career during the interbellum period, it was appropriate to facilitate their extinction through "birth-control regulation, segregation, or human sterilization"; soon thereafter he cited "the Negro" as an example of the "lower mental capacity" that warranted such treatment. While such observations may have reflected the eugenics era's zeitgeist, well into the 1970s he was still promoting the concept of "genthanasia" for "phasing out" a less capable group through "educational and birth control measures." Though he now refrained from naming names, his reference to a group with low intelligence but resistance to malaria provided an unmistakable hint for those aware that many people of African descent carry the sickle cell gene, which confers malarial immunity.[74]

In addition to the cognitive elite's greater value financially, civically, and as progenitors of future generations, IQ scientists have also fostered a view of their lives as worth more in some deeper fundamental sense, one that violates traditional beliefs about the equal value of all lives just by virtue of their humanity. And because the lives of the cognitive elite matter more in this basic sense, the loss of their lives is considered a greater source of concern than the loss of their intellectual inferiors. When the philosopher Michael Scriven noted that "the worth of people and their rights do not depend on IQ," for example, Shockley disagreed because, he claimed, test scores were correlated with other important traits, suggesting that human worth was indeed predicated on intelligence to some degree. And although the proposal in Stanley Kubrick's classic film *Dr. Strangelove* was intended as black comedy—that in case of nuclear war persons with high IQs be selected for survival in underground shelters—Shockley was not joking when he suggested that nuclear war might serve as a "grim possibility" for solving "the problem of the quality of the human race" by forcing society to select the most intelligent from among the survivors to perpetuate life on the planet. As always Cattell did not bother with subtleties. To clinch the case that persons of different intelligence consequently differed in their innate value, he offered what he regarded as such an obvious and compelling example that it required no comment: "as if a motorist in an unavoidable situation would hesitate to run over ... a feebleminded in preference to a healthy, bright child."[75] An IQ score is thus regarded more as a verdict than a measurement, a judgment of the value of one's life.

Although Herrnstein and Murray ended *The Bell Curve* on a high note, emphasizing the importance of governmental policies that enable "people to live lives of dignity" no matter their intelligence, earlier in the concluding chapter they too hinted at the greater importance of the cognitive elite to the polity and of their children to the future welfare of the country. They didn't suggest that anyone be deprived of the franchise, but they did find it essential that government remain the province of the "natural aristocracy." They didn't call for anyone to be involuntarily sterilized, but they did offer a eugenic rationale for their policy recommendation straight out of nineteenth-century Social Darwinism. A "society with a higher mean IQ is also likely to be a society with fewer social ills and brighter economic prospects," wrote Herrnstein and Murray, and "the most efficient way to raise the IQ of a society is for smarter women to have higher birth rates than duller women." But too many poor women, "disproportionately at the low end of the intelligence distribution," were having children, portending "a future America with more social ills and gloomier economic prospects." Thus, they argued, in words that could have been written by Herbert Spencer, providing financial assistance of any kind to the poor "subsidizes births" among those "who are also disproportionately at the low end of the intelligence distribution," only "encouraging the wrong women" to reproduce, perpetuating their genetic deficiencies, and thereby undermining the intellectual level of the nation as a whole.[76] And after half a thousand pages of data supposedly demonstrating that IQ scores were the major factor associated with more favorable outcomes for just about every meaningful social variable—income, employment, education, criminal behavior, health, infant mortality, and more—it was hard to escape the implication that the cognitive elite were just more valuable as human beings.

This notion that the lives of the more intelligent have greater innate worth has become a widespread meme as exemplified in a *New York Times* article that appeared in June 1995. Headlined "Sudden End for 2 Who Had Everything to Live For," the 900-word article described in some detail the lives of two persons who had been gunned down at lunchtime in midtown Manhattan, both of them likely members of the cognitive elite: a computer graphics designer with a degree in communications from the University of Wisconsin "bursting with creative energy" and a Phi Beta Kappa graduate from Saint Olaf with a master's degree in architecture from Yale and "a long list of clients." An accompanying box with two-sentence sketches of each victim indicated, however, that seven people altogether had been killed in the rampage, the other five occupying mundane positions that apparently did not qualify them for the condition in the headline: a cab driver, a parking lot attendant, a blackjack dealer, a market company owner suspected of drug dealing, and the mother of the killer's ex-girlfriend.[77] Murray himself may have found the headline objectionable, but there is no doubt that it reflected *The Bell Curve*'s subtext as well as the thinking of numerous IQ researchers who preceded Herrnstein and Murray.

Although there may be a few specific contexts that justifiably require monetary calculation of an individual life's value—settlement of an insurance claim for wrongful death, for example—the belief that some lives are intrinsically more valuable than others should be morally offensive. Notwithstanding the forced choice posed by Cattell, the death of someone with a low IQ is not ipso facto less grievous than the death of someone more cognitively capable. As the philosopher K. Anthony

Appiah reminds us, every person, no matter their abilities, faces the "challenge of making a meaningful life. The lives of the less successful are not less worthy than those of others–but not because they are *as* worth or *more* worthy. There is simply no sensible way of comparing the worth of human lives."[78] Even the "very dull," those whom *The Bell Curve* characterized as soon to become "a net drag," incapable of "putting more into the world than they take out," have "everything to live for."

Of course, this is not to deny that there are crucial positions in the society requiring talent, education, and effort, and it is important to identify the people whose attributes make them most capable of fulfilling these roles and to provide material incentives encouraging them to do so. But quite apart from the question of whether the resulting income inequality is justifiable, differences in the inherent valuation of lives— in the esteem, dignity, and respect accorded to people—can be no less important as a source of human motivation. Indeed, it was these latter differences that played the more significant role in the 2016 presidential election.

Notes

1. R.J. Herrnstein, "In Defense of Bird Brains," *The Atlantic Monthly* (September 1965): 101–104. C. Ingraham, "Forget Robots—The Goats are Coming for Our Jobs," *Washington Post*, July 7, 2017.
2. R.J. Herrnstein, *I.Q. in the Meritocracy* (Boston: Little, Brown, 1973), 6–7; see also R.J. Herrnstein, "On Challenging an Orthodoxy," *Commentary* (April 1973): 53. A.R. Jensen, "How Much Can We Boost IQ and Scholastic Achievement," *Harvard Educational Review* 39 (1969): 1–123.
3. G. Piel, "…Ye May Be Mistaken," in *Genetic Destiny*, ed. E. Tobach and H.M. Proshansky (New York: AMS Press, 1976), 132. B. Rice, "The High Cost of Thinking the Unthinkable," *Psychology Today* (December 1973): 91.
4. R.J. Herrnstein, "I.Q." *The Atlantic Monthly* 228 (September 1971): 43–64. Herrnstein, *I.Q. in the Meritocracy*, 53. The controversy over the *Atlantic* article eventually brought new attention to the twin study cited by Herrnstein as the largest of its kind and the only one with data

affirming the crucial assumption of no relation between the occupational status of the homes in which the twin pairs were raised; as a result the study was demonstrated to be scientifically worthless and probably fraudulent. See W.H. Tucker, "Fact and Fiction in the Discovery of Sir Cyril Burt's Flaws," *Journal of the History of the Behavioral Sciences* 30 (1994): 335–347 and W.H. Tucker, "Re-reconsidering Burt: Beyond a Reasonable Doubt," *Journal of the History of the Behavioral Sciences* 33 (1997): 145–162.

5. Herrnstein, "I.Q." 62–64. F. Galton, *English Men of Science: Their Nature and Nurture* (London: Macmillan, 1874), 23.
6. Herrnstein, "I.Q." 63–64. Caploe, "Herrnstein in 'The Atlantic' Predicts American Meritocracy."
7. Herrnstein, "I.Q." 63. Herrnstein, *I.Q. in the Meritocracy*, 215. M. Young, *The Rise of the Meritocracy* (New Brunswick, New Jersey: Transaction Publishers, 1958). M. Young, "Down with Meritocracy," *Guardian*, June 28, 2001. University of North Carolina sociologist Bruce K. Eckland, too, took the book seriously, referring to "Young's meritocracy [in which] talented adults rise to the top of the social hierarchy and the dull fall or remain at the bottom"; see B.K. Eckland, "Genetics and Sociology: A Reconsideration," *American Sociological Review* 32 (1967): 181.
8. E. Barker, *The Politics of Aristotle* (London: Oxford University Press), 17. A. Bloom, *The Republic of Plato* (New York: Basic Books, 1968), 94.
9. D.G. Ritchie, *Natural Rights: A Criticism of Some Political and Ethical Conceptions* (New York: Macmillan, 1903), 258. F.H. Hankins, "Individual Differences and Democratic Theory," *Political Science Quarterly* 38 (1923): 409.
10. E.G. Boring, "Lewis Madison Terman: 1877–1956, *Biographical Memoirs of the National Academy of Sciences*, 33 (1959): 414. L.M. Terman, *The Measurement of Intelligence* (Cambridge, Massachusetts: Riverside Press, 1916), 91–92. L.M. Terman, "The Significance of Intelligence Tests for Mental Hygiene," *Journal of Psycho-Asthenics* 18 (1914): 124. L.M. Terman, *The Intelligence of School Children* (Boston: Houghton Mifflin, 1919), 270, 288.
11. C. Burt, "Psychological Tests for Scholarship and Promotion," *School* 13 (1925): 741. C. Burt, "Individual Psychology and Social Work," *Charity Organization Review* 43 (1918): 18. On the posthumous exposure of Burt's research see supra, note 28.

12. C. Spearman, *The Abilities of Man* (New York: Macmillan, 1927), 8.
13. Young, *The Rise of the Meritocracy*, 85–87.
14. D.R. Caploe, "Herrnstein in 'The Atlantic' Predicts American Meritocracy," *Harvard Crimson*, September 22, 1971. Herrnstein, *I.Q. in the Meritocracy*, 200–201.
15. Herrnstein and Murray *The Bell Curve*, 520.
16. Herrnstein and Murray *The Bell Curve*, 443, 442. R. Spencer, "The Charlottesville Statement" appears on the alt-right website, https://altright.com/2017/08/11/what-it-means-to-be-alt-right/.
17. Herrnstein and Murray *The Bell Curve*, 518, 523, 526.
18. Ibid., 528–532.
19. Ibid., 541–545.
20. Murray is one of five writers contributing to I.M. Stelzer, "The Shape of Things to Come," *National Review* (July 8, 1991): 30. C. Goldin and R. Margo, "The Great Compression: The Wage Structure in the United States at Mid-Century," *Quarterly Journal of Economics* 107 (1992): 1–34. Analysis of U.S. Census Bureau data in L. Mishel and J. Bernstein, *The State of Working America 1994–95* (M.E. Sharpe, 1994), 37. The decline from 48 to 20 percent is cited in R. Frank, *Richistan: A Journey Through the American Wealth Boom and the Lives of the New Rich* (New York: Three Rivers, 2007), 39. D. Bell, "On Meritocracy and Equality," *Public Interest* 29 (1972): 64.
21. See "Trends in the Distribution of Income," CBO Blog, October 25, 2011, https://www.cbo.gov/publication/42537. E. Saez and G. Zucman, "Alexandria Ocasio-Cortez's Tax Hike Idea Is Not About Soaking the Rich," *New York Times*, January 22, 2019. D.C. Johnston, "Richest Are Leaving Even the Rich Far Behind," *New York Times*, June 5, 2005, 1.
22. E. Saez and G. Zucman, "Wealth Inequality in the United States Since 1913: Evidence from Capitalized Income Tax Data," *NBER Working Paper Series*, http://gabriel-zucman.eu/files/SaezZucman2014.pdf. C. Collins and J. Hoxie, *Billionaire Bonanza: The Forbes 400 and the Rest of Us* (Washington, DC: Institute for Policy Studies, November 2017), 2, 4. E. Sherman, "America is the Richest, and Most Unequal, Country," *Fortune*, September 30, 2015, http://fortune.com/2015/09/30/america-wealth-inequality/. R. Menon, "The United States Has a National-Security Problem—And It's Not What You Think," *Nation*, July 16, 2018, https://www.thenation.com/article/united-states-national-security-problem-not-think/.

23. T. Piketty, *Capital in the Twenty-first Century* (Cambridge, Mass.: Belknap/Harvard University Press, 2014), 264, 265.
24. D. Markovits, *The Meritocracy Trap: How America's Foundational myth Feeds Inequality, Dismantles the Middle Class, and Devours the Elite* (New York: Penguin, 2019), 5, 98.
25. Ibid., 105–106.
26. On the accurate story behind Hemingway's misquoted comment, see E. Dow, "The rich Are Different," letter to the editor, *New York Times*, November 13, 1988, 70. P. Fussell, *Class: A Guide Through the American Status System* (New York: Summit Books, 1983), 29–30. N.D. Schwartz, *The Velvet Rope Economy: How Inequality Became Bug Business* (New York: Doubleday, 2020), 16, 4. On the coronavirus vaccine, see S. Kahn, "How Rich People Will Cut the Line for the Coronavirus Vaccine," *Washington Post*, December 18, 2020.
27. Herrnstein, *I.Q. in the Meritocracy*, 124; also Herrnstein, I.Q., 51.
28. Herrnstein, I.Q., 64.
29. Ibid., 51, 63. N. Chomsky, "Psychology and Ideology," *Cognition: International Journal of Cognitive Psychology* 1 (1972): 39. For the complete list of occupations by IQ, see Herrnstein, *I.Q. in the Meritocracy*, 120–121. R.J. Herrnstein, "Whatever Happened to Vaudeville? A Reply to Professor Chomsky," *Cognition: International Journal of Cognitive Psychology* 1 (1972): 303.
30. S. Brill, *Tailspin: The People and Forces Behind America's Fifty-Year Fall—And Those Fighting to Reverse It* (New York: Alfred A. Knopf, 2018), 26. L. Uchitelle, "Lure of Great Wealth Affects Career Choices," *New York Times*, https://www.nytimes.com/2006/11/27/business/27richer.html. Markovits, *The Meritocracy Trap*. 239.
31. P. Krugman, "Gilded Once More," *New York Times*, April 27, 2007, A27. "Top 25 Highest Paid Hedge Fund Managers of 2008," March 26, 2009, https://www.marketfolly.com/2009/03/top-25-highest-paid-hedge-fund-managers.html. P. Krugman, "Bernie Sanders and the myth of the 1 Percent," *New York Times*, April 18, 2019, A25. G. Thompson, "Meet the Wealth Gap," *Nation*, June 30, 2008, 20. S. Polk, "For the Love of Money," *New York Times*, January 19, 2014, SR1.
32. A. Griswold, "Harvard Grads Are Still Flocking to Finance," *Moneybox: A Blog About Business and Economics*, May 27, 2014, www.slate.com/blogs/moneybox/2014/05/27/harvard_class_of_2014_elite_grads_are_still_flocking_to_finance_and_consulting.html; W.F. Morris IV, "Har-

vard's Wall Street Problem," *Crimson*, February 17, 2016, https://www.thecrimson.com/article/2016/2/17/-Wall-Street-Problem-Morris/. C. Rampell, "Out of Harvard and Into Finance," *New York Times*, December 11, 2011, https://economix.blogs.nytimes.com/2011/12/21/out-of-harvard-and-into-finance/. C. Simon, "Prestige Draws Penn Students to Finance and Consulting after Graduation—But at What Cost?" *Daily Pennsylvanian*, April 26, 2017, https://www.thedp.com/article/2017/04/consulting-finance-popularity.

33. A. Roe, "A Psychologist Examines 64 Eminent Scientists," *Scientific American* 187 (1952): 22, 25. C. Murray, *In Pursuit of Happiness and Good Government* (New York: Simon and Schuster, 1988), 235, 238. Markovitz, *The Meritocracy Trap*, 40, 192–194.
34. J. Rothwell, "Myths of the 1 Percent: What Puts Some People at the Top," *New York Times*, November 24, 2017, B2.
35. See H.R. Gold, "Never Mind the 1 Percent. Let's Talk About the .01 Percent," *Chicago Booth Review*, Winter 2017/18, http://review.chicagobooth.edu/economics/2017/article/never-mind-1-percent-lets-talk-about-001-percent. Markovitz, *The Meritocracy Trap*, 164. The comment by Carnegie was cited in J. Madrick, "How to Succeed in Business." *New York Review of Books*, (April 18, 1996): 22. W.D. Cohan, "Lehman's Demise, Dissected," *New York Times*, March 18, 2010, https://opinionator.blogs.nytimes.com/2010/03/18/lehmans-demise-dissected/.
36. D. Bok, *The Cost of Talent: How Executives and Professionals Are Paid and How It Affects America* (New York: Free Press, 1993), 238.
37. W.D. Cohan, "When Lenders Push Borrowers Over the Edge," *New York Times*, May 13, 2019, A19. L. Story and E. Dash, "Bankers Reaped Lavish Bonuses During Bailouts," *New York Times*, July 30, 2009, A1. See also B. Whiteman, "What Red Ink? Wall Street Paid Hefty Bonuses," *New York Times*, January 28, 2009, A1. S. Anderson, *Off the Deep End: The Wall Street Bonus Pool and Low-Wage Workers* (Washington: Institute for Policy Studies, March 8, 2016), 2.
38. B. Covert, "Everyone Must Go," *Nation* 308 (May 6, 2019): 23. E. Applebaum and R. Batt, "Private Equity Pillage: Grocery Stores and Workers at Risk," *American Prospect* (Fall, 2018). P. Whoriskey and D. Keating, "Overdoses, Bedsores, Broken Bones: What Happened When a Private-Equity Firm Sought to Care for Society's Most Vulnerable," *Washington Post*, November 25, 2018. M. Corkery and B. Protess,

"How the Twinkie Made the Superrich Even Richer," *New York Times*, December 11, 2016, A1.
39. S, Pizzigatti, *The Case for a Maximum Wage* (Cambridge, UK: Polity, 2018).
40. W.G. Sumner (ed. By A.G. Keller), *The Challenge of Facts and Other Essays* (New Haven: Yale University Press, 1914), 90. Thorndike, *Human Nature and the Social Order*, 95. C. Wright Mills, *The Power Elite* (New York: Oxford University Press, 2000), 118.
41. Mills, *The Power Elite*, 116, 129. T. Cowen, "CEOs Are Not Overpaid," *Time*, April 22, 2019, 22.
42. B. Saporito, "C.E.O. Pay, America's Economic 'Miracle'" *New York Times*, May 17, 2019, https://www.nytimes.com/2019/05/17/opinion/ceo-pay-raises.html. Morgan is cited in G.S. Crystal, *In Search of Excess: The Overcompensation of American Executives* (New York: W. W. Norton, 1991), 24. P.F. Drucker, "Reform Executive Pay or Congress Will," *Wall Street Journal*, April 24, 1984, 34. Drucker's letter is quoted in J. McGregor, "What's the Right Ratio for CEO-to-Worker Pay?" *Washington Post*, September 19, 2013. I. Salisbury, "The Average CEO Makes as Much Money in One Day as the Typical Worker Earns in a Full Year, *Time*, May 22, 2018, http://money.com/money/5287123/ceo-pay-afl-cio/. D. Gelles, "Where the Buck Doesn't Stop," *New York Times*, May 27, 2018, BU6.
43. See the Wikipedia entry for Charles Erwin Wilson. Markovitz, *The Meritocracy Trap*, 183, 176.
44. Crystal, *In Search of Excess*, 11, 31. Bok, *The Cost of Talent*, 16, 111, 100.
45. L. Bebchuk and J. Fried, *Pay Without Performance: The Unfulfilled Promise of Executive Compensation* (Cambridge: Harvard University Press, 2004), chapter 8.
46. Bok, *The Cost of Talent*, 117–118.
47. Bebchuk and Fried, *Pay Without Performance*, 133. "Ford CEO: $28M for 4 Months Work," April 5, 2007, https://money.cnn.com/2007/04/05/news/companies/ford_execpay/. J. Creswell and M. Barbaro, "Home Depot Board Ousts Chief, Saying Goodbye With Big Check, *New York Times*, January 4, 2007, A1. M. Huckman, "Pfizer's McKinnell: The $200 Million Dollar Man," December 22, 2006, https://www.cnbc.com/id/16326224. See the Wikipedia entry on Hewlett Packard: https://en.wikipedia.org/wiki/Hewlett-Packard. D. Gelles, "Muilenburg, Fired C.E.O. Will Receive More Than $60 Million," *New York Times*,

January 10, 2020. C. Reinicke, "Boeing sees $11 Million of Market Value Erased in Just 2 Days as Its 737 MAX Disaster Worsens," *Markets Insider*, December 17, 2019, https://markets.businessinsider.com/news/stocks/boeing-stock-price-falls-erases-billions-2-days-737-max-halt-2019-12-1028769301.
48. Bebchuk and Fried, *Pay Without Performance*, 181.
49. A. Bhattarai and D. Santamariña, "Bonuses Before Bankruptcy: Companies Doled Out Millions to Executives Before Filing for Chapter 11, *Washington Post*, October 26, 2020.
50. Mills, *The Power Elite*, 56, 131.
51. Brill, *Tailspin*, 29–31, 56. Markovitz, *The Meritocracy Trap*, 11, 184.
52. Ibid., 111–114.
53. Bok, *The Cost of Talent*, 240.
54. Brill, *Tailspin*, 106, 110.
55. E. Lipton and K.P. Vogel, "Fine Print of Stimulus Package, Special Deals for Certain Industries," *New York Times*, March 6, 2020, A8. A. Abramson, "Federal Stimulus Spending is Giving the lobbying Industry a Giant Windfall," *Time*, May 2, 2020. Quoted in J. Drucker, "Bonanza Hides in a Rescue Package," *New York Times*, April 25, 2020, B1.
56. Markovitz, *The Meritocracy Trap*, 54, 58. J. Eisinger, J. Ernsthausen, and P. Kiel, "The Secret IRS Files: Trove of Never-Before-Seen Records Reveal How the Wealthiest Avoid Income Tax, ProPublica, June 8, 2021, https://www.propublica.org/article/the-secret-irs-files-trove-of-never-before-seen-records-reveal-how-the-wealthiest-avoid-income-tax.
57. For details of their offenses and sentences, see the Wikipedia pages for each of the executives. J. Eisinger, *The Chickenshit Club: Why the Justice Department Fails to Prosecute Executives* (New York: Simon & Schuster, 2017), 318.
58. Ibid., 199–200, xix.
59. Markovitz, *The Meritocracy Trap*, 267.
60. See the transcript of Kelleher's appearance on "The Beat with Ari Melber," March 9, 2020, http://www.msnbc.com/transcripts/msnbc-live-with-ari-melber/2020-03-09.
61. R. Niebuhr, *Moral Man and Immoral Society: A Study in Ethics and Politics* (New York: Charles Scribner's Sons, 1932), 8. Herrnstein, I.Q., 53.
62. A.F. Rubalcava, "Recruit This, McKinsey," *Harvard Crimson*, November 26, 2001, https://www.thecrimson.com/article/2001/11/26/recruit-this-mckinsey-times-are-tough/.

63. K. Ho, *Liquidated: An Ethnography of Wall Street* (Durham, N.C.: Duke University Press, 2009), 39–40, 76. Markovitz, *The Meritocracy Trap*, 168.
64. L.A. Rivera, "Ivies, Extracurriculars, and Exclusion: Elite Employers' Use of Educational Credentials," *Research in Social Stratification and Mobility* 29 (2011): 72–73, 78–80. See also L.A. Rivera, *Pedigree: How Elite Students Get Elite Jobs* (Princeton, N.J.: Princeton University Press, 2015).
65. N. Barofsky, *Bailout: An Inside Account of How Washington Abandoned Main Street While Rescuing Wall Street* (New York: Free Press, 2012), 138, 139.
66. H.H. Goddard, *Human Efficiency and Levels of Intelligence* (Princeton, N.J.: Princeton University Press, 1920), vi, 100–101. Thorndike, *Human Nature and the Social Order*, 370–372.
67. Spearman, *The Abilities of Man*, 8. Goddard, *Human Efficiency and Levels of Intelligence*, 99. H.H. Goddard, *Psychology of the Normal and Subnormal*, 234. L.M. Terman, *The Measurement of Intelligence* (Cambridge, Mass.: Riverside, 1916), 91–92. L.M. Terman, "The Psychological Determinist; or Democracy and the IQ," *Journal of Educational Psychology* 6 (1922): 62.
68. G.B. Cutten, "The Reconstruction of Democracy," *School and Society* 16 (1922): 478–481. W. McDougall, *Ethics and Some Modern World Problems* (London: Methuen, 1925), 156–163.
69. R.B. Cattell, *The Fight for Our National Intelligence* (London: P.S. King, 1937), 59, 109. R.B. Cattell, *Beyondism: Religion from Science* (New York: Praeger, 1987), 113, 114, 223, 224. R.P. Oliver, *Christianity and the Survival of the West* (Cape Canaveral: Howard Allen, 1973), 75.
70. F. Galton, "hereditary Improvement," *Fraser's Magazine* 7 (1873): 129. Spearman, *The Abilities of Man*, 8. On the history, see P. Reilly, *The Surgical Solution: A History of Involuntary Sterilization in the United States* (Baltimore: Johns Hopkins University Press, 1991).
71. D.J. Ingle, *I Went to See the Elephant* (New York: Vantage, 1963), 213. D.J. Ingle, "Genetic Bases of Individuality and of Social Problems," *Zygon: Journal of Religion and Science* 6 (1971): 182, 189. D.J. Ingle, *Who Should Have Children?* (New York: Bobbs-Merrill, 1973), 102, 115.
72. See Shockley's address as a Nobel Laureate in *Genetics and the Future of Man*, ed. J.D. Roslansky (New York: Appleton-Century-Crofts, 1965), 67. W. Shockley, "Dysgenics, Geneticity, Raceology: A Challenge to the Intellectual Responsibility of Educators," *Phi Delta Kappan* 53,

special supplement (January 1972): 306. On the suggestion of "bounty hunters," see A.R.S. Goodell, *The Visible Scientists* (Stanford: Proquest Dissertations Publishing, 1975), 339–340. J. Fincher, "Arthur Jensen: In the Eye of the Storm," *Human Behavior* (March/April 1972): 22. Jensen, "How Much Can We Boost IQ and Scholastic Achievement," 95.

73. D.J. Ingle, letter from the editor, *Perspectives in Biology and Medicine* 11 (1968): 713. W. Shockley, letter to the editor, *Scientific American* 224 (January 1971): 6. Jensen, "How Much Can We Boost IQ and Scholastic Achievement," 95.

74. R.B. Cattell, *Psychology and Social Progress: Mankind and Destiny from the Standpoint of a Scientist* (London: C.W. Daniel, 1933), 317, 322, 323, 360. R.B. Cattell, *The Fight for Our National Intelligence* (London: P.S. King, 1937), 56. R.B. Cattell, *A New Morality from Science: Beyondism* (New York: Pergamon, 1972), 153–154, 221.

75. M. Scriven, "The Values of the Academy," *Review of Educational Research* 40 (1971): 546. W. Shockley, "Negro IQ Deficit: Failure of a 'Malicious Coincidence' Model Warrants New Research Proposals," *Review of Educational Research* 41 (1971): 243. "Is Quality of U.S. Population Declining? Interview With a Nobel Prize-winning Scientist," *U.S. News & World Report*, November 22, 1965, 71. Cattell, *The Fight for Our National Intelligence*, 67–68.

76. Herrnstein and Murray, *The Bell Curve*, 530, 548, 551.

77. C. Goldberg, "Sudden End for 2 Who Had Everything to Live For," *New York Times*, June 23, 1995, B2.

78. On efforts to assign monetary value to a life, see H.S. Friedman, *Ultimate Price: The Value We Place on Life* (Berkeley: University of California Press). K.A. Appiah, "The Red Baron," *New York Review of Books* (October 11, 2018): 23.

Open Access This chapter is licensed under the terms of the Creative Commons Attribution 4.0 International License (http://creativecommons.org/licenses/by/4.0/), which permits use, sharing, adaptation, distribution and reproduction in any medium or format, as long as you give appropriate credit to the original author(s) and the source, provide a link to the Creative Commons license and indicate if changes were made.

The images or other third party material in this chapter are included in the chapter's Creative Commons license, unless indicated otherwise in a credit line to the material. If material is not included in the chapter's Creative Commons license and your intended use is not permitted by statutory regulation or exceeds the permitted use, you will need to obtain permission directly from the copyright holder.

3

Politics and Intelligence: Running Against the Cognitive Elite

While Herrnstein and Murray believed firmly in genetic inequality, which, they argued, both explained and justified social and economic inequality, they also vigorously supported *political* equality; indeed, they suggested that humans could not be equal "in any other sense." Citing the beliefs of the founding fathers as support, they asserted that "the best government was one that most efficiently brought the natural aristocracy to high positions." And they expressed confidence that the "common people" had the good sense to choose what Madison called "men of virtue and wisdom" to govern—that is, those members of the cognitive elite prepared for such a role by their natural ability and their broad education in "history, literature, arts, ethics, and the sciences." The great majority of citizens—that 95 percent not as intelligent as the cognitive elite—might not possess the right characteristics for the governing class, but, according to *The Bell Curve*, they could be counted on to recognize those who did.[1]

In 2016, the Democratic Party nominated a candidate for President whose background as an epitomical member of the cognitive elite closely matched The *Bell Curve*'s description of the natural aristocrat destined for political leadership. Graduate of a prestigious liberal arts college where she received the institution's highest academic honor, the first student ever to address the school's commencement—a speech covered by the *New York Times* and reprinted in *Life Magazine*—research assistant for the House Republican caucus the summer before her senior year, both research assistant for a seminal text on child custody and a volunteer providing legal services to the poor while completing her law degree at Yale, author of a highly cited article on children's rights in *Harvard Educational Review*, all before going on to serve as First Lady, Senator from New York, and Secretary of State, Hillary Rodham Clinton would seem to embody that combination of "education of a particular kind" and commitment to public service which, according to Herrnstein and Murray, marked her as fit to govern.[2] Whatever her undeniable shortcomings—the sense of political calculation that seemed to inform so many of her decisions, the careless treatment of her emails, the questionable futures contracts that paid off so handsomely, the inept public statements after the Benghazi attack—they would seem to pale in comparison with the flaws of her opponent: a notoriously thin-skinned, narcissistic, louche, reality-television host and shady real-estate wheeler-dealer in his third marriage with a penchant for childish, petty insults, a history of half a dozen bankruptcies despite having inherited more than 400 million dollars from his father, and a well-documented record for swindling people; in addition, Donald Trump had been sued thousands of times, insulted war heroes, mocked the disabled, attributed a hostile question from a journalist to her menstrual period, bragged of sexually assaulting women and barging into the dressing rooms of teen-age beauty pageants to leer at the contestants, referred to avoiding sexually transmitted diseases as his personal equivalent of Vietnam, and, in probably the crassest moment in the history of presidential politics, actually touted the size of his penis in a debate with other Republican hopefuls. Hillary Clinton enjoyed a well-deserved reputation as a policy wonk; before announcing his candidacy for the nation's highest office, Donald Trump had no record of public service, and his main involvement in any issue of public or political significance was to spearhead the campaign to delegitimize the first African-American President. From *The Bell Curve*'s viewpoint, the outcome of the election was startling to say the least.

Trump's victory clearly had little to do with either a coherent ideology or an enduring set of principles or policies, mainly because he had no firmly held beliefs of any kind, typically favoring whatever seemed most beneficial to his self-interest at a particular moment; basking in the adulation of the crowds at his rallies seemed to take priority over attachment to any particular substantive position. Over the years, he had changed party affiliation at least five times, ranging from Manhattan liberal, who had donated to the campaigns of Anthony Weiner, Andrew Cuomo, Elliot Spitzer, Chuck Schumer, Kamala Harris and Hillary Clinton, to right-wing zealot, even registering for a couple of years as a member of the Independence Party; pronounced himself "very pro-choice" before embracing the pro-life cause; argued that women should be punished for having an abortion at the same time that he praised Planned Parenthood; advocated for a single-payer universal healthcare system before rejecting it as a government takeover; supported a ban on assault rifles and longer waiting periods for gun purchases before opposing all such measures as a champion of the Second Amendment; had a long history of employing undocumented workers on his building sites and at his golf clubs before making illegal immigration the centerpiece of his candidacy; first supported and then opposed military action in Afghanistan, Iraq, and Libya, swinging easily, as a group of Republican national security experts pointed out, "from isolationism to military adventurism within the space of one sentence"[3]; had at various times both opposed and supported an increase in the minimum wage, which he first thought should be set by the federal government and then decided should be left to the states; and donated more than one hundred thousand dollars to the Clinton Foundation before denouncing it during a presidential debate as a "criminal enterprise."

Nor could Trump's election be attributed to a sense that the nation was experiencing a crisis of some kind or heading in the wrong direction. In the fall of 2016, the United States was undeniably the world's most powerful nation; by that time, its unemployment rate had returned from the double digits at the height of the Great Recession to just under the five percent figure regarded by many economists as the defining point for "full employment" and the poverty rate was near a historic low; the rate of major crimes—murder, "forcible rape," robbery, and aggravated assault—was similarly at or near historic lows; and, for all the technical difficulties that plagued the onset of the Affordable Care Act, "Obamacare" had reduced the uninsured share of the population by half. While economic problems certainly remained and should not be minimized—especially the decline of the manufacturing sector and its effect on working families—little at the time could justify Trump's melodramatic pronouncement of "American carnage."

The 2016 election split the country in large part across educational lines. According to the statistician Nate Silver's analysis, Clinton won handily in the fifty most well-educated counties in the nation, while Trump similarly prevailed in the fifty least. In Berkeley, home of the nation's most prestigious public university, Clinton received ninety percent of the vote, while Trump, with three percent, finished third, behind the Green Party candidate. White men without college degrees voted for the Republican candidate at the highest rate since exit polls began; when Trump famously proclaimed his love for "the poorly educated," it was not without ample justification. Among journalists, the disparity was unprecedented: of those newspapers and magazines that endorsed one of the major party candidates, 406 chose Clinton, including many traditionally conservative publications that had not preferred a Democrat in decades; 26 supported Trump, only two of which had circulation greater than 100,000. Clinton received endorsements from 77 college newspapers, Trump none. In a geographical reflection of this educational divide, Trump carried those areas well outside the city center that still relied substantially on manufacturing—the blue-collar workers who had constituted the Democratic base half a century earlier—while the Democrats were now what a Stanford political scientist called "the party of urban, postindustrial America."[4]

In classic populist fashion, Trump exploited this educational division to create a narrative of conflict between a privileged, parasitic elite, undeserving of its position, and the common folk—between a highly educated but aloof class of people who exercised power to their own advantage as multicultural, global citizens, and the real Americans, a silent majority who sensed instinctively what was right for the country without having to rely on "expert"—i.e., elite—advice. As Trump himself expressed it in the *Wall Street Journal* six months before his election, "The only antidote to ruinous rule by a small handful of elites is a bold infusion of popular will. On every major issue affecting this country, the people are right and the governing elite are wrong."[5] People in general in this rhetoric were not meant to be synonymous with "*the* people," only some of the former qualifying for inclusion in the latter.

Thus, rather than campaigning on any substantive agenda, Trump ran as a representative of an aggrieved minority resentful of the worldview espoused by *The Bell Curve*, in which differences in intelligence are offered as justification not only for income inequality but for differences in social status. Neither the book nor its remaining author was mentioned during the campaign, and Trump's frequent reference to the "elite" was never preceded by the word "cognitive"; indeed, given his limited range of information, it is possible that the future president was not even aware of the academic controversy. But it was clear that this notion of an elite—a "natural aristocracy," as Herrnstein and Murray had put it—entitled by intellect and education to its prerogatives, provided Trump with a foil against which he posed as avatar of the rage of those average people who sensed their exclusion from this favored group; Trump's policies might not help them, but he hated the same people they did. Instead of recognizing the "hoi aristoi," as *The Bell Curve* had predicted, the "common people" apparently resented them, and the fact that Trump's opponent, clearly considering herself a member of the elite, characterized so many of his supporters as "deplorables" only served to confirm these feelings of resentment on their part.

As an increasingly "woke" society focused its concern on the plight of "marginalized" groups, working-class Whites, who were largely excluded from this sympathetic rubric, were told—despite their undeniable personal struggles, the loss of their manufacturing jobs in a globalized economy, and the consequent decimation of their communities—that they benefitted from a "privilege" based on their skin color. The title of journalist Ben Bradlee Jr.'s book—an in-depth exploration of the thinking of Trump supporters in one Pennsylvania county—accurately summarized its conclusion: "The Forgotten." In an influential essay on "The Politics of Recognition," published around the same time as *The*

Bell Curve, the Canadian political philosopher Charles Taylor maintained that "Due recognition ...is a vital human need," and "misrecognition shows not just a lack of due respect. It can inflict a grievous wound." While Taylor wrote with a different context in mind—"minority ... groups, some forms of feminism and what is ... called the politics of 'multiculturalism'"—his analysis seemed particularly appropriate to the Trump campaign's recognition of working-class, non-college-educated Whites as people whose lives mattered.[6] The oxymoronic notion of a populist billionaire became the vehicle for converting their pain into pride.

Of course, it was the liberals who had a lengthy history of overt opposition to Murray's views, but they did not regard his conclusions as an adverse comment on their own abilities; rather, their outrage reflected indignation at the Platonic rigidity of *The Bell Curve*'s claim that a genetic characteristic exerted such a determinative effect on a person's life. For Trump supporters, however, their obvious hostility at any mention of the elite had nothing to do with abstract concepts of a fair society; it was intensely personal. In an obvious allusion to the Godfather's feckless son, Fredo, the conservative activist and former speechwriter for George W. Bush, David Frum, described their thinking as "I'm not dumb, I'm smart, and I want respect." These were people who knew they were smart, even if not in the same way as those effete country-club smart-asses with their ability to express themselves more articulately and their greater degree of cultural literacy. As the *New York Times* columnist David Brooks described it, the elites encouraged an image of themselves as "enlightened" people, who had attended "competitive colleges, ... have the brainpower to run society and who might just be a little better than other people." Trump punctured that impression, offering a counternarrative in which these "meritocrats are actually clueless idiots and full of drivel, and ... virtue, wisdom and toughness is found in the regular people whom those folks look down upon."[7] The elites in this view might be comfortable handling a nine-iron but were clueless about how to use a tire iron.

Despite his own wealth, Trump was the ideal vessel to channel this sense of grievance, being inclined by both nature and experience to play the victim; "I have been treated very badly" was his mantra both before and after becoming president. Trump had long been looked down on by the elites in his home city as a crass and tasteless arriviste from an outerborough, unwelcome in Manhattan society no matter how much money he had inherited or how many buildings he erected. To begin construction on his eponymous tower he jackhammered the historically significant Art Deco bas-relief sculptures adorning the Bonwit building his tower had replaced, even though the Metropolitan Museum of Art had expressed interest in obtaining them; calling the act "esthetic vandalism," the *New York Times* noted that "big buildings do not make big human beings." Then, shortly after the tower's opening—with its ostentatious atrium, exclusive shops on the lower floors and luxury condos above, purchased in many cases by celebrities—the *Times* called it "pretentious," noting that his critics viewed Trump as "a raving egomaniac" and "a rogue billionaire, loose in the city like some sort of movie monster, unrestrained by the bounds of good taste." Other New York journalists dismissed him as "a bridge-and-tunnel guy," whom sophisticated Manhattanites "laughed at" and considered "repulsive." In perhaps the most personally cutting incident, at a black-tie dinner attended by prominent entertainers and journalists, Trump found himself the object of mockery by the suave and urbane president he would seek to replace. Nor was it only liberal intellectuals who regarded Trump as tacky and boorish. The "Never Trumpers," prominent conservative members of the cognitive elite originally attracted to the movement by the thought of people like William F. Buckley and Irving Kristol, bristled at the prospect of this yahoo as their standard bearer. George Will, for example, responded to Trump's announcement of his intent to run by calling him "an unprecedentedly and incorrigibly vulgar presidential candidate" responsible for the "coarsening of civic life."[8] Trump knew what it felt like to be dismissed as an uncultured rube and, even though extremely rich, could identify with that sense of resentment felt by the adoring crowds that flocked to his campaign events, eager to give the finger to the cognitive elite; their grievances were his grievances.

This conflict between the elites and the common people became a central theme of both the Trump campaign and the Trump administration, regularly invoked at the rallies held by the President both before and after his election. In a typical riff, Trump derided the "people they call the elite" as "stone cold losers," declaring that "I have a better education than them. I'm smarter than them. I went to the best schools. ... Much more beautiful house, much more beautiful apartment, much more beautiful everything. And I'm President, and they're not." It was his supporters, the people at his rally, Trump proclaimed, who truly merited the designation: "You're smarter," he told them; "You're sharper. You're more loyal. You're the elite. We're the elite." And when he called on them to defend "your dignity" and take back "your country," no doubt existed about the identity of the villains who were responsible for the assault on their dignity and had usurped control of the nation from its rightful owners: it was the people *The Bell Curve* had labeled the "cognitive elite."[9]

Thus, the political irony: although Murray, a prominent fellow at one of the centers of conservative thought, was despised by progressives for his unrelenting opposition to programs of government support as well as his conclusions about racial differences in intelligence, nevertheless Trump ran as essentially the anti-Murray. *The Bell Curve* had argued that the cognitive elite rose to their appropriate position at the top of the class structure and in the halls of power because they are smarter than others; Trump reassured those who had been left behind that they struggled not because they lacked intelligence because they had been betrayed by the people with the high IQs. That this tactic proved successful should perhaps not be surprising. The meritocratic elite, whose wisdom informed policy, had instituted the financial rules that led to the Great Recession, presided over the global flow of capital that had decimated so much of American manufacturing, and provided the rationale for blundering into one disastrous campaign after another for regime change in the Middle East. When candidate Trump famously responded to a question about his policy advisors by announcing that "my primary consultant is myself," it didn't seem all that bizarre in view of the previous decade and a half.

For all his anti-elitist rhetoric, however, Trump generally refrained from attacking the uber-rich together with their corporate and legal enablers, a task that fell to the other populist in the 2016 race: Bernie Sanders, too, presented himself as a champion of the people against the powerful, though he rarely referred to the latter as "elites," preferring to characterize them as the "millionaires and billionaires," who had profited from financialization of the economy at everyone else's expense. In contrast, Trump considered the extraordinarily wealthy his peer group—or at least wished to foster that perception in others. Shortly after the election, Trump announced that "I want people who made a fortune" in the cabinet—he would then appoint the richest cabinet in American history—and a billionaires' row enjoyed the best seats at the inauguration, sitting onstage with the President. According to an investigation by the political journalist Robert Draper, President Trump "enjoyed being around billionaires"—people that one administrative official described as "superrich guys who wouldn't give him the time of day" before the election. The new President also "stocked his Intelligence Advisory Board with wealthy businesspeople," who would discomfit intelligence officials by asking questions related to their business interests; the chair of the Advisory Board, for example, was the co-executive of the private equity firm that owned a major defense contractor with a number of military contracts. And after being lobbied by a number of billionaire investors—Sheldon Adelson, John Paulsen, Thomas Barrack,, and others—Trump pardoned Michael Milken, the junk bond king and 1980s symbol of greed, convicted of various counts of securities and tax fraud.[10] After all, these were the kind of people who purchased memberships in Mar-a-Lago, and rented office space and bought condos in his buildings.

Trump's anti-elitism thus turned out to be highly selective, eschewing criticism of the finance professionals and private equity moguls, whose self-proclaimed brilliance had wreaked such havoc on the economy, and focusing his outrage instead on the knowledge elite, that segment of the cognitive elite that had chosen to pursue careers in the sort of professions crucial to the functioning of a complex, modern society—specialists with deep knowledge in their area. American democracy had long been marked by a distrust of so-called "eggheads" and a belief that the native wisdom of common people was preferable to the unrealistic assessments of intellectuals; when Richard Nixon, as Dwight Eisenhower's running mate in 1952, called Democratic presidential nominee Adlai Stevenson an "egghead," it contributed to the impression that he was out of touch with ordinary citizens. And William F. Buckley, himself an intellectual, famously declared that he would "rather be governed by the first 2,000 people in the Boston telephone directory than by the 2,000 people on the faculty of Harvard University." But this suspicion of educated sophisticates reflected a hostility to the prospect of their political leadership, not the factual information they could provide. In contrast, Trump's opposition expressed itself as contempt for knowledge; as the Yale historian Beverly Gage observed, "Trump has taken it to a whole new level by not only attacking clueless elites but the entire idea of expertise." In fact, according to the journalist Michael Lewis, immediately after the election, when the Obama administration organized the traditional briefings designed to ensure a smooth transition of power—for every federal agency a team of 30–40 people from the outgoing president meets with a similar group from the incoming in order to explain the working of the department—"the Trump people weren't anywhere to be found"; not believing they had anything to learn, they just didn't show up—not even for the session with the Department of Energy, an agency headed by a nuclear physicist on leave from MIT and charged with responsibility for the nation's nuclear arsenal.[11] Essentially, Trump ran and then governed against those with professional expertise, characterizing them as conspirators threatening the people's sovereignty.

A study conducted during the run-up to the 2016 election by two political scientists at the University of Minnesota provided empirical support for the appeal of this approach to Trump's base. A nationally representative sample of more than one thousand adult American citizens responded to a battery of questions about their opinions of the political process, many of the items adapted from survey studies of populism. The responses were subjected to principal components analysis, yielding three uncorrelated dimensions. The first dimension reflected "feelings of marginalization relative to wealth and political power," characterized by agreement with such statements as "It doesn't really matter who you vote for because the rich control both political parties"; the authors named it "anti-elitism." As one would expect, supporters of the two populist candidates scored the highest on this dimension with Sanders's people slightly ahead of Trump's; none of the supporters of the other major candidates at the time (Hillary Clinton, Ben Carson, Ted Cruz, Marco Rubio, John Kasich) came close. The second dimension indicated "a general skepticism of science and expert opinion," exemplified by agreement with statements such as "I'd rather put my trust in the wisdom of ordinary people than the opinions of experts and intellectuals" and "When it comes to really important questions, scientific facts don't help very much." On this dimension, which the authors called "mistrust of expertise," Trump's supporters scored the highest and Sanders's the lowest by a large margin, only Clinton's supporters coming anywhere near as low. That is, the anti-elitism expressed by Sanders's supporters was rooted in economic inequality and what they perceived as the resulting political powerlessness, but it was not at all conjoined to the extreme distrust of expertise that characterized the Trump supporters. On the third dimension, "national affiliation," reflecting a "collectivist 'American' identity," Trump supporters again scored the highest and Sanders supporters the lowest, suggesting a populism that was not only anti-elitist but also ethno-nationalist. Echoing Sarah Palin's reference during the 2008 campaign to specific parts of the country as "the real America," this emphasis on a connection to national greatness helped to provide a basis for an otherwise lacking social status.[12]

Daniel Markovitz, the Yale law professor, sees such pervasive mistrust of expertise as an unavoidable consequence of meritocracy. "Because meritocracy identifies skill and expertise with elites," he writes, "it condemns middle-class workers who accept the value of knowledge and training to internalizing their own exclusion and degradation," requiring a rejection of expertise as a form of self-respect. As an expression of this logic, he points out, class resentment is directed not so much at entrepreneurs, even when very wealthy, but at professionals—those members of the cognitive elite whose status is based not on their incomes but on their education and knowledge. As Joan C. Williams, the legal scholar at the University of California Hastings College of Law and founding director of the Center for WorkLife Law, observes, blue-collar workers admire the rich, with whom they have little direct contact, but resent professionals because "professionals order them around every day."[13]

Trump transformed this antipathy to professional expertise from rhetoric to reality, especially concerning issues not just involving abstract policy but affecting people's decisions and behaviors in ways perceived by the President and his supporters as condescending attempts by the cognitive elites to determine how others, presumably less knowledgeable than themselves, should live their lives—decrees by sheltered academics and bureaucratic smart-asses about what kind of straws and light bulbs people must use; the economist Paul Krugman calls such opposition "regulation rage," coming from people who "don't feel respected, and who see even mild restrictions on their actions as insults perpetrated by elites who consider themselves smarter than other people."[14] It was no accident that so many Trump supporters chose as their emblem the Gadsden flag, its "Don't tread on me" motto an announcement of their refusal to submit to illegitimate authority.

Nowhere did this rejection of professional expertise have more impact than Trump's response to the novel coronavirus, which he treated as a wedge issue, providing another opportunity to attack elites. The President refused to endorse wholeheartedly the behavioral recommendations from public health officials designed to minimize the virus, instead offering encouragement to those supporters who chose to flaunt their disregard of these measures. But, just as Krugman had noted, this opposition was informed, not so much by a sense that the directives—to wear a mask, to avoid unnecessary travel, etc.—were ineffective, but rather by resentment at the presumptuousness, the effrontery, of the elites who assumed they could tell others how to behave. The voices on Fox News regularly echoed the anger of Trump's supporters at being told what to do. Tucker Carlson, for example, host of the most watched program in the history of cable news as of the end of 2020, referred contemptuously to the lead member of the White House Coronavirus Task Force as "Lord Fauci" and told his viewers that the doctor's "strategy" was to "keep ordering you around like an animal" so that you wouldn't notice that he was "totally incompetent." Before the 2020 holiday season, Fauci regularly appeared on television, pleading with people to stay home and avoid interaction with those outside their immediate household. "We hope you ignore that advice," Carlson bluntly told his viewers. It wasn't about the pandemic, he insisted; it was about "social control." The goal of the elites, he declared, was to make it "a crime to live a normal life." Especially for those struggling economically, resistance, first to the covid restrictions urged by scientists, and then to vaccination signified personal dignity and independence; after all, if expertise is tyranny, then the refusal to abide by its pronouncements becomes a declaration of freedom.

Although this hostility to expertise associated with elites exerted a particularly harmful effect on personal responses to the coronavirus, it also played a less noticed role in the strange subordination of data to the Trump administration's wishful thinking. That is, the normal sequence, in which statements are based on evidence, was reversed, requiring that the latter be manufactured to provide support for Trump's preference for the former; if those with expertise refused to compromise their integrity to produce the desired result, others without qualifications were solicited. For example, when the model created by epidemiologists projected a number of deaths from the coronavirus considered "too catastrophic," someone in Jared Kushner's office with a background in finance was ordered to create an alternative, receiving explicit instructions for the obligatory conclusion: "They [the epidemiologists] think 250,000 people could die" and this model should "show that fewer than 100,000 people will die in the worst case scenario," the finance specialist was told. Public health experts at the Centers for Disease Control and Prevention were ordered to rewrite guidelines for reopening schools that Trump found too stringent, and when the President decided that extensive testing was resulting in too many cases of the virus, the CDC issued new testing guidelines, not written by agency scientists but imposed from above, that were called alarming and dangerous by medical experts. In a joint statement, four prior directors of the CDC, having served under both Democratic and Republican Presidents, referred to "these repeated efforts to subvert sound public health guidelines" as "putting lives at risk"; they could "not recall over our collective tenure a single time when political pressure led to a change in the interpretation of scientific evidence." Perhaps worst of all, senior CDC officials complained that the agency's most highly respected publication on infectious disease, "The Morbidity and Mortality Weekly Reports," was being turned into "a political loyalty test, with career scientists framed as adversaries of the administration."[15]

Prominent scientists disinclined to affirm Trump's erroneous assertions quickly found their role diminished in favor of more politically reliable spokespersons, characterized by an editorial in the prestigious

New England Journal of Medicine—in a departure from its two-century-long practice of refraining from political comment—as "charlatans who obscure the truth and facilitate the promulgation of outright lies." (*Scientific American*, too, broke its 175-year tradition of political neutrality and endorsed Joe Biden in 2020, noting that Trump's "rejection of evidence and public health measures have been catastrophic.") One of the country's premier experts on vaccine development, the director of the Biomedical Advanced Research and Development Authority (the Department of Health and Human Service office responsible for medical countermeasures against bioterrorism and pandemic diseases) was fired after opposing the funding for what he described as "potentially dangerous drugs promoted by those with political connections and by the administration itself." And when the director of the CDC's National Center for Immunization and Respiratory Diseases warned of severe disruption to everyday life during a White House press briefing, she was replaced by scientists who could be relied on to phrase their assessments with less candor.[16] It was as if a pilot warned that Air Force One was unsafe and the President insisted on a different pilot; elites with expertise were welcome in the Trump administration only if they would tell the President what he already wanted to believe.

Murray himself emerged as one of Trump's harshest critics, appalled that "we should have a malignant, narcissist grifter with dementia in the presidency," though the blame for the election of such an unfit individual to the nation's highest office he now placed squarely on the same group that he had once pronounced certain to "lead the country no matter what": the cognitive elite. In Murray's account of the process that had led to their culpability, when "the nation's most prestigious colleges and universities," which had once been the province of "bluebloods and the wealthy," were opened to "youth from all backgrounds" who could demonstrate "talent, pluck and hard work," the new policy did not produce the "socioeconomic democratization" he had expected. At first, some young people not from privileged backgrounds benefited from such opportunities, he noted, but this "phase lasted only a generation or two," as these supersmart youth of both sexes, segregated from the less capable and destined for economic success in a society in which "brains have become radically more valuable in the marketplace," married each other, thereby "combining their large incomes and genius genes" to "produce offspring who get the benefit of both." Consequently, "isolated from mainstream America and ignorant about the lives of ordinary Americans", this genetically favored group developed "a distinctive culture," differing substantially from the majority of their fellow citizens—in "the food they eat, the way they take care of their health, their child-rearing practices, the vacations they take, the books they read, the websites they visit"—and resulting in their "condescension toward," and even "contempt for ordinary Americans."[17]

It was this arrogance on the part of the cognitive elite that Trump had successfully campaigned against, Murray argued, apparently oblivious to the role that his own work had played in fostering the syndrome he now condemned. *The Bell Curve* had described how, more than a quarter century before Trump's candidacy, the cognitive elite—the "cream floating on the surface of American society"—had begun to constitute its own class, the members of which had little in common with their fellow citizens. Herrnstein and Murray then addressed their readers in a chummy style that took for granted that they were all members of this select group; this was clearly a book written *by* the cognitive elite *for* the cognitive elite. "Most readers of this book," they declared, "are in preposterously unlikely groups": many of their dozen closest friends are not only college graduates but hold advanced degrees—a result less likely than one in a million in a randomly selected group. And then, changing entirely to the second person, they wrote that "You—meaning the self-selected person who has read this far into this book—live in a world that probably looks nothing like" The Bell Curve. "In all likelihood, almost all of your friends and professional associates belong" to the cognitive elite, while those "whom you consider to be unusually slow are probably somewhere in Class II"—the 20 percent of the curve just below the cognitive elite but still well above average. That is, *The Bell Curve* congratulated its readers on the fact that their social world was so highly selective that even their "unusually slow" friends were smarter than 75 percent of the population. And shortly after publication of *The Bell Curve*, Murray explained to a journalist that in the past people were poor because of bad luck or social barriers; but now, he declared, "what's holding them back is that they're not bright enough to be a physician."[18]

Indeed, although Murray clearly sympathized with the common folk's hostility to what he called the "New Elite," even his description of the latter's emergence contained an unsubtle reminder of the former's shortcomings. After all, if, as he noted, those elite institutions that provided the gateway to success were no longer just the province of the wealthy and the well-connected but now accessible to anyone with the requisite drive and ability, then the implication was unavoidable: others might resent the "people who count" but were excluded from their ranks only by their own intellectual or personal deficiencies. Thus, despite his condemnation of elite snobbery, both *The Bell Curve* and Murray's own subsequent pronouncements had helped to lay the groundwork for its existence. If elitist condescension did in fact play a role in Trump's election, it was an attitude that Murray had done his part to promote. Having earlier helped to poison the well, Murray then sought to present himself as the water inspector.

Notes

1. Herrnstein and Murray *The Bell Curve*, 444, 531.
2. F.M. Hechinger, "Mood of Campus Valedictory Speakers Is Somber," *New York Times*, June 15, 1969, 54. "The Class of '69," *Life Magazine* 66 (June 20, 1969): 31. The text on child custody was J. Goldstein, A. Freud and A.J. Solnit, *Beyond the Best Interests of the Child* (New York: Free Press, 1973). H. Rodham, "Children Under the Law," *Harvard Educational Review* 43 (1973): 487–514.
3. See the statement "coordinated" by E.A. Cohen and B. McGrath, signed by 122 "members of the Republican national security community," "Open Letter on Donald Trump from GOP National Security Leaders," https://warontherocks.com/2016/03/open-letter-on-donald-trump-from-gop-national-security-leaders/.
4. N. Silver, "Education, Not Income, Predicted Who Would Vote for Trump," *FiveThirtyEight*, https://fivethirtyeight.com/features/education-not-income-predicted-who-would-vote-for-trump/. On endorsements, see the Wikipedia entry, "Newspaper Endorsements in the 2016 United States Presidential Election, https://en.wikipedia.org/wiki/Newspaper_endorsements_in_the_2016_United_States_presidential_election. J.A.

Rodden, *Why Cities Lose: The Deep Roots of the Urban–Rural Political Divide* (New York: Basic Books, 2019), 42.

5. D.J. Trump, "Let Me Ask America a Question," *Wall Street Journal*, April 15, 2016.
6. B. Bradlee Jr., *The Forgotten: How the People of One Pennsylvania County Elected Donald and Changed America* (New York: Little, Brown, 2018). C. Taylor, "The Politics of Recognition," in *Multiculturalism: Examining the Politics of Recognition*, ed. A. Gutmann (Princeton: Princeton University Press, 1994), 25, 26.
7. Frum is quoted in R.P. Saldin and S.M. Teles, *Never Trump: The Revolt of the Conservative Elites* (New York: Oxford University Press, 2020), 175. D. Brooks, "When Politics Becomes Your Idol," *New York Times*, October 23, 2017, A23.
8. "Crumbling Patrimony," *New York Times*, June 15, 1980, 20E. W.E. Geist, "The Expanding Empire of Donald Trump," *New York Times Magazine*, April 8, 1984, 30. See the observations by Lou Colasuonno, former editor of the *New York Post* and the *New York Daily News* and by Dan Rather, quoted in M. Kruse, "How Gotham Gave Us Trump," *Politico* (July/August 2017), https://www.politico.com/magazine/story/2017/06/30/donald-trump-new-york-city-crime-1970s-1980s-215316. G.F. Will, "Donald Trump Is a Counterfeit Republican," *Washington Post*, August 12, 2015.
9. Quoted in, for example, B. Cole, "Donald Trump on Liberal Elites: 'I'm Smarter then Them. I Went to the Best Schools,' Says He Has 'A Much More Beautiful House'," *Newsweek*. March 29, 2019, https://www.newsweek.com/donald-trump-liberal-elites-smarter-schools-beautiful-house-1379578.
10. K.P. Vogel and M. Conway, "Trump on Cabinet: 'I want people that made a fortune'," *Politico*, December 8, 2016, https://www.politico.com/story/2016/12/trump-cabinet-billionaires-fortune-232403. D. Alexander, *White House, Inc.: How Donald Trump Turned the Presidency Into a Business* (New York: Portfolio/Penguin, 2020), 65–66. R. Draper, "Unwanted Truths: Inside Trump's Battles With U.S. Intelligence Agencies," *New York Times Magazine*, August 8, 2020, 32. J.B. Stewart and J. Drucker, "Milken's Ties Spurred Bid for Pardon," *New York Times*, March 2, 2020, B1.
11. On Wallace, see T.H. Anderson, "The 1968 Election and the Demise of Liberalism," *South Central Review* 34 (2017): 44. Gage is quoted in

Quoted in P. Wehner, *The Death of Politics* (New York: HarperCollins, 2019), 201. M. Lewis, *The Fifth Risk* (New York: Norton, 2018), 35–37.

12. J.E. Oliver and W.M. Rahn, "Rise of the *Trumpenvolk*: Populism in the 2016 Election," *Annals of the American Academy of Political and Social Science* 667 (2016): 189–206.
13. Markovitz, *The Meritocracy Trap*, 64. J.C. Williams, "What So Many People Don't Get About the U.S. Working Class," *Harvard Business Review* (web article), November 10, 2016, https://hbr.org/2016/11/what-so-many-people-dont-get-about-the-u-s-working-class.
14. P. Krugman, "The Roots of Regulation Rage," *New York Times*, September 18, 2019.
15. Y. Abutaleb, A. Parker, J. Dawsey and P. Rucker, "The Inside Story of How Trump's Denial, Mismanagement and Magical Thinking Led to the Pandemic's Dark Winter," *Washington Post*, December 19, 2020. L.H. Sun and J. Dawsey, "CDC Feels Pressure from Trump as Rift Grows Over Coronavirus Response," *Washington Post*, July 9, 2020. S.G. Stolberg, "Top U.S. Officials Told C.D.C. to Soften Coronavirus Testing Guidelines," *New York Times*, August 27, 2020. T. Frieden, J. Koplan, D. Satcher and R. Besserd, "We Ran the CDC. No President Ever Politicized its Science the Way Trump Does," *Washington Post*, July 14, 2020. N. Weiland, S.G. Stolberg and A. Goodnough, "Political Appointees Meddled in CDC's 'Holiest of the Holy' Health Reports," *New York Times*, September 12, 2020.
16. "Dying in a Leadership Vacuum," editorial, *New England Journal of Medicine* 383 (2020): 1480. "From Fear to Hope," editorial, *Scientific American* 323 (2020): 12–13. The Director of BARDA, Rick Bright, is quoted in M/ Goldberg, "We're All Casualties of Trump's War on Science," *New York Times*, May 11, 2020, 26. G. Lopez, "How Trump Let Covid-19 Win," *Vox*, August 24, 2020, https://www.vox.com/future-perfect/21366624/trump-covid-coronavirus-pandemic-failure.
17. Quoted in Saldin and Teles, *Never Trump*, 176. C. Murray, "The Tea Party Warns of a New Elite. They're Right," *Washington Post*, October 24, 2010. C. Murray, "Trump's America," *Wall Street Journal*, February 12, 2016. J. Miltimore, "Charles Murray Interview: On Trump, the Chaos at Middlebury, and America's Greatest Threat," *Intellectual Takeout*, June 7, 2017, https://www.intellectualtakeout.org/article/charles-murray-interview-trump-chaos-middlebury-and-americas-greatest-threat/.

18. Herrnstein and Murray, *The Bell Curve*, 47, 121. Murray is quoted in DeParle, "Daring Research or 'Social Science Pornography'?" 48.

Open Access This chapter is licensed under the terms of the Creative Commons Attribution 4.0 International License (http://creativecommons.org/licenses/by/4.0/), which permits use, sharing, adaptation, distribution and reproduction in any medium or format, as long as you give appropriate credit to the original author(s) and the source, provide a link to the Creative Commons license and indicate if changes were made.

The images or other third party material in this chapter are included in the chapter's Creative Commons license, unless indicated otherwise in a credit line to the material. If material is not included in the chapter's Creative Commons license and your intended use is not permitted by statutory regulation or exceeds the permitted use, you will need to obtain permission directly from the copyright holder.

4

Conclusion: Addressing Inequality

Although it was Herrnstein's writing in the early 1970s that began widespread discussion of the concept, and *The Bell Curve*, which he co-authored with Murray that continued it, the truth is that the notion of a "meritocracy" has become an abiding conviction across much of the political spectrum in the United States. The belief that society's "winners" deserve their hugely disproportionate share of resources because they are better—i.e., smarter—than others is not unique to conservatives and libertarians like Murray; it is also an article of faith for much of the so-called "New Democratic" establishment that has controlled the party from the Clinton through the Obama administrations. For both conservatives and many liberals, the meritocratic faith is not so much a way to explain inequality as to rationalize it; high-ranking officials involved in economic policy in both Republican and Democratic administrations have considered inequality not only inevitable but the appropriate reflection of people's economic value. As John Snow, George W. Bush's Treasury Secretary, bluntly put it, "people will get paid on how valuable they are to the enterprise." And Obama's first Director of the National Economic Council, Larry Summers, who had earlier served as Clinton's Treasury Secretary, declared that inequality had increased only because "people are being treated closer to the way that they're supposed to be treated."[1] (In his former capacity, however, Summers had fought to deregulate Wall Street, presiding over enactment of the Financial Services Modernization Act—the "Graham-Leach-Bliley" act—which, by removing the Glass-Steagall safeguards on banks, led to the Great Recession, which did so much to exacerbate inequality.)

Ultimately, then, *The Bell Curve* constituted a natural ideological traveling companion for the neoliberalist consensus dominating policy circles across party lines, providing data-based support for a radically individualist agenda seeking to crush organized labor, drive down tax rates on the wealthy, deregulate business, and privatize much of the public sector. Four decades of this approach have transformed the American economy, resulting in a minimalist welfare state, an erosion of the relationship between social role and financial reward and a dramatic increase in inequality.

Neoliberalism is based on two interrelated assumptions. First, it reflects a view of the society and the economy, in which the market functions rationally, selecting for greater income and status those who deserve their position by virtue of talent, natural ability, or hard work. As a consequence, inequality becomes solely the result of individual differences, whether in specific abilities, amount of education, or various personality traits; *The Bell Curve*'s insistence on intelligence as the single most important such variable represented merely a special case of this argument. Indeed, in an even more extreme version, not long after the book's publication Murray found it "almost certainly" the case that the poor were burdened by a "genetic makeup that is significantly different" from the configuration in the rest of the population.[2] This focus on individual differences, some of them viewed as unalterable, implies a view of dramatic inequality as fair in some sense. Some people do well, while others do not, but in either case they get what they deserve, based on the market's determination of their value. Success is due to those with the right constellation of traits, while failure becomes a consequence of personal defect or character flaw for others, who are lacking in some fundamental respect—intelligence, ambition, drive, determination. And just as Michael Young predicted, the former, who have no doubt that their status has been earned and is thus deserved, have reason to be proud, while the latter feel humiliated, all the more so as increasing equality of opportunity deprives them of any explanation for their plight other than their own shortcomings.

The second assumption, almost a corollary of the first, is best encapsulated by Margaret Thatcher's famous declaration that "there's no such thing as society. There are individual men and women and there are families. And no government can do anything except through people, and people must look after themselves." Public institutions, in this view, are inappropriate mechanisms for responding to private difficulties. *The Bell Curve*'s argument converted Thatcher's claim of contradiction from the abstract to the empirical. After all, if IQ scores accounted for much of the variation in individual outcomes—not only economic success but a host of other variables related to quality of life—then social democratic reforms and other collective protections for the less fortunate could not change what was essentially a state of nature: innate human differences.

This emphasis on individual differences as both explanation and justification for massive economic inequality hinders our ability to understand and respond to more significant sources of the problem. Indeed, social psychologists studying how people behave in specific contexts refer to the tendency to over-emphasize the importance of personal traits and underestimate the role of situational factors as the "fundamental attribution error"; rather than being a characteristically impatient Type A personality, for example, the person speeding past other cars in traffic may be responding to a genuine emergency. However, this often-erroneous focus solely on individual characteristics as the determining factor for behavior is no less an attributional error in explaining the kind of more enduring life outcomes studied by *The Bell Curve*, where the more significant variable is not some ephemeral situational element but the underlying social structure—the rules of the society and the economy. Given the history of discriminatory practices in the United States, this omission is particularly egregious in any attempt to explain racial differences in income and wealth.

Of course, individual differences in drive and ability account for some portion of the variation in people's economic outcomes, but they pale in comparison with the explanatory power of structural variables. To cite but one example, consider the effect of the decline in unionism.

Through the first three decades after the war, a thriving union movement was instrumental in helping to create the world's largest middle class. At one point, more than a third of workers in the private sector enjoyed the benefits of collective bargaining; at present that figure is 6.3 percent. And representation produced gains for non-union workers as well, putting pressure on employers to improve their treatment, lest the company face an organizing drive. A 2014 study by three economists from non-profit institutes called it "possible to explain the entire rise of inequality since the late 1970s as the outcome of an array of economic policies," with de-unionization alone as the cause of "a third of the entire growth of wage inequality among men."[3]

In a parallel analysis, half a century ago the psychologist William Ryan distinguished between two approaches to social problems. The "exceptionalist" approach assumes that there are "specially defined categories of persons," who "'have' social problems as a result of some kind of unusual circumstances—accident, illness, personal defect or handicap, character flaw or maladjustment—that exclude them from using the ordinary mechanisms for maintaining and advancing themselves." Even a quite common problem in this view suggests only a large number of instances of these individual deviances. Especially on issues such as income or health, this approach "concentrates almost exclusively on the failure of the deviant," the ways in which, as a result of individual imperfections, the person who "has" the problem is unable to adapt appropriately to the circumstances or the system. And if the problem originated as a result of individual defect, then the remedy too had to be individualistic, tailored to the specific case. Thus, the exceptionalist approach, which Ryan called "Blaming the Victim," led to solutions that were "private, voluntary, remedial, special, local, and exclusive"; the person with the problem had to be "fixed."[4]

In contrast, what Ryan called the "universalist" approach attributes social problems to "defects in the community and the environment rather than in the individual." This view "see[s] social problems, in a word, as social"—as a "function of the social arrangements of the community or the society." And since, he argued, these arrangements were often "quite imperfect and inequitable, such problems are both predictable and, more important, preventable through public action."

Thus, the universalist approach implied solutions "that are public, legislated, promotive or preventive, general, national, and inclusive," focusing not on individuals but on "the development of standard generalized programs affecting total groups."[5] Curing pica—the tendency for some young children to consume harmful substances like lead paint chips—is exceptionalist; enforcing the housing laws that prohibit the use of lead-based paint in residential facilities is universalist. Also universalist is a program like unemployment insurance, which protects people from the vagaries of the business cycle, while carrying no implication that those individuals who benefit from it are defective or abnormal.

To the extent that neoliberalism considers extreme inequality a problem to be ameliorated rather than an inescapable reality to be embraced, its preferred solutions are all exceptionalist, focusing exclusively on the presumptive flaws of those who have struggled economically: they don't have *enough* education; they need a different *kind* of education; they need a different skill set for the postindustrial economy; their cognitive skills (i.e., IQ) must be improved. Even the emphasis on equal opportunity as a response to inequality is ultimately an exceptionalist approach. As the Harvard philosophy professor Michael J. Sandel notes, "Enabling people to compete solely on the basis of effort and talent would bring market outcomes into alignment with merit," ensuring that people receive what the market determines as "their just deserts"; such an emphasis accepts the fact "that the rungs on the income ladder were growing farther apart" and seeks merely "to help people compete more fairly to clamber up the rungs," thereby exacerbating rather than reducing inequality.[6] This is not to derogate the importance of attempts either to improve people's abilities or to dismantle arbitrary barriers to achievement like race, class, or gender; both of these are worthy goals for practical as well as moral reasons. But such exceptionalist responses to what the sociologist Tressie McMillan Cottom calls "the 'skillification' of the U.S. economy" will do little to reduce inequality resulting from structures and institutions; not everyone can become what Robert Reich called a "symbol analyst." And to tell people who are struggling after their factory was moved abroad that their salvation lies in more education, which will better enable them to compete, is more of a provocation than a solution.[7]

A universalist attempt to address inequality would focus on programmatic measures designed to ensure that everyone who makes a contribution to the common good can be assured of certain essentials necessary for a decent quality of life: income sufficient to provide reasonable shelter and food, access to appropriate health care and education, and the ability to retire with dignity. Although this is not the place to explore the kind of policies that would achieve such goals, there has been no shortage of proposals for doing so: substantially increasing the minimum wage, a negative income tax, a universal basic income (UBI), some form of universal health care, a federally supported, guaranteed retirement account, etc. Indeed, 58 mayors, including those from 6 of the 10 largest cities in the United States, have joined Mayors for a Guaranteed Income, an organization "based on the truth that financial instability is not the failure of individuals, but rather policies." The schemes for funding these proposals typically shift the tax burden from work to wealth; Sandel, for example, suggests raising revenue through a "financial transaction tax on high-frequency trading, which contributes little to the real economy."[8] Numerous Western European democracies, no less industrialized or technologically advanced than the United States, have managed to provide such benefits and even more, such as parental leave and a minimum number of vacation days, undeterred by the inevitable range of personal characteristics in their populations.

Long after *The Bell Curve*, Murray himself announced his support for a modest UBI for people at the lower end of the income spectrum, though only under a draconian condition that would leave most recipients worse off with this benefit than without it. In exchange for a $3000 grant to be applied specifically toward the cost of health insurance and then a monthly payment of $833 dollars (i.e., $10,000 per year), every other form of federal assistance would be terminated: "Social Security, Medicare, Medicaid, food stamps, Supplemental Security Income, housing subsidies, welfare for single women and every other kind of welfare and social-services program." The fact that, as a result of eliminating these programs, "the wealth in private hands would be greater than ever before" Murray considered an advantage, leading, he anticipated, to "restoration, on an unprecedented scale, of a great American tradition of voluntary efforts to meet human needs."[9] In place of universalist measures to address inequality, Murray offered the ultimate example of the exceptionalist approach: widespread reliance on private charity.

The pandemic has provided a painful reminder of the gap between what the market rewards and what actually contributes to the public weal. It is not just the fact that the wealthy have enjoyed enormous financial gains since spring 2020, while so many others have suffered: the 15 richest Americans have added more than 400 billion dollars to their net worth, the total wealth owned by American billionaires has grown 55 percent, and a survey of CEOs "revealed some of the biggest pay packages on record," even as their companies laid off thousands of employees, forcing them to turn to government assistance for food and housing. But in addition, the pandemic has produced the sudden realization that the tasks performed by so many modestly compensated workers—care givers, cleaners, transportation workers, waste collectors, drug store workers, grocery store clerks, delivery workers, and others— make essential contributions, without which the society could not function. In a particularly cruel irony, the meat- and poultry-processing industry lobbied the Trump administration, successfully, to have its line workers—composed substantially of immigrant labor, much of it undocumented and all of it miserably paid—classified as "essential," ensuring that the plants remained open, even as covid spread rapidly through the ranks of people required to work elbow-to-elbow even while lacking health insurance; a number of governors enacted measures granting these companies immunity from civil liability, protecting them from any claim for compensation filed by the families of those who died after workplace exposure to the virus.[10]

It is not only logically contradictory but morally offensive to claim that certain activities are so vital to the functioning of the society that the workers who engage in them should endure risks from which others are shielded, while at the same time disparaging what they do as so "low-skilled" that they are not entitled to decent wages, based on the market's rational analysis of the value of their efforts. And to tell these "essential" workers and so many others in traditional working-class jobs–people who may not be members of the cognitive elite but who produce the truly useful goods and services in the economy—that their economic struggles result from a lack of intelligence or education is insulting. They don't need better jobs; they need better compensation and more respect for the underpaid and underappreciated jobs they already have.

While some degree of inequality in a free society is to be expected, the extreme mismatch between social contribution and financial remuneration in the United States is the result of policy decisions, not the inevitable consequence of individual differences, whether or not genetic. As *The Bell Curve*'s subtitle—"Intelligence and Class Structure in American Life"—indicates, the book sought to argue otherwise, promoting an emphasis on individual cognitive differences that effectively thwarts consideration of any real solutions to the problem of inequality. That is its greatest danger.

Notes

1. On Snow, see Editorial Board, "Let's Talk About Higher Wages," *New York Times*, November 28, 2020, SR8. Summers is quoted in R. Suskind, *Confidence Men: Wall Street, Washington, and the Education of a President* (New York: HarperCollins, 2011), 197.
2. C. Murray, "Deeper into the Brain," *National Review*, January 24, 2000, 48.
3. "Union Members—2020," Bureau of Labor Statistics, January 22, 2021. L. Mishel, J. Schmitt and H. Shierholz, "Wage Inequality: A Story of Policy Choices," *New Labor Forum* (August 4, 2014): 1, 3, https://files.epi.org/charts/wage-inequality-a-story-of-policy-choices.pdf.
4. W. Ryan, *Blaming the Victim* (New York: Vintage, 1971), 14–17.
5. Ibid.
6. M.J. Sandel, *The Tyranny of Merit: What's Become of the Common Good?* (New York: Farrar, Straus and Giroux, 2020), 63, 85.
7. T.M. Cottom, *Lower Education: The Troubling Rise of For-Profit Colleges in the New Economy* (New York: New Press, 2017), 20. R. Reich, *The Work of Nations: Preparing Ourselves for 21st Century Capitalism* (New York: Knopf, 1992).
8. For example, N. Zewde, K. Strickland, K. Capatosto, A. Glogower, and D. Hamilton, *A Guaranteed Income for the 21st Century* (New York: The New School Institute on Race and Political Economy, May 2021);

T. Ghilarducci and T. James, *Rescuing Retirement: A Plan to Guarantee Retirement Security for All Americans* (New York: Columbia University Press, 2020). M. Carter, E. Garcetti and M. Tubbs, "Most Americans Support Guaranteed Income," *Time*, July 8, 2021; for a list of the mayors, see https://www.mayorsforagi.org/. Sandel, *The Tyranny of Merit*, 219.

9. C. Murray, "A Guaranteed Income for Every American," *Wall Street Journal*, June 3, 2016.
10. On gains by the wealthy, see D. Markovitz, "We Need to Tax Wealth," *Time*, May10/May 17, 2021, 28; H. Olen, "As wealthy CEOs rake in money, an ugly trope about Americans needing help reemerges," *Washington Post*, May 19, 2021; and P. Eavis, "They're Cashing In Like Never Before," *New York Times*, June 13, 2021, BU1. A. Driver, "Their Lives on the Line," *New York Review of Books*, April 29, 2021, https://www.nybooks.com/daily/2021/04/27/their-lives-on-the-line/.

Open Access This chapter is licensed under the terms of the Creative Commons Attribution 4.0 International License (http://creativecommons.org/licenses/by/4.0/), which permits use, sharing, adaptation, distribution and reproduction in any medium or format, as long as you give appropriate credit to the original author(s) and the source, provide a link to the Creative Commons license and indicate if changes were made.

The images or other third party material in this chapter are included in the chapter's Creative Commons license, unless indicated otherwise in a credit line to the material. If material is not included in the chapter's Creative Commons license and your intended use is not permitted by statutory regulation or exceeds the permitted use, you will need to obtain permission directly from the copyright holder.

Index

A
abortion 83
   Planned Parenthood 83
   pro-choice campaign 83
   pro-life cause 83
abstract reasoning 64
accountants 39
Adams, John 33
Adelson, Sheldon 89
affirmative action 12–15, 34
Affordable Care Act (US) 84
Afghanistan
   military action in 83
agnosticism 6, 7
AIG insurance 63
altruism 11
American Enterprise Institute 1
American Lawyer, The 54
American Renaissance 11, 12

American Sociological Association 27
animal learning 22
anti-elitism 89–91
Apotheker, Leo 52
Appiah, K. Anthony 71
Arabic literature 7
aristocracy 25, 27, 37
   genetic 24
   natural 32, 33, 69, 81, 85
Aristotle 27, 28, 31
   government by hoi aristoi 27, 85
art 7
Art Deco 87
Asians 11
assault
   aggravated 84
   sexual 82
Atlantic Monthly, The 21
Atlantic, The 21–24, 26, 30, 71–73

attorneys 53

B
Bankers Trust 62
banking. *See also* financial industry
  global system 52
Banking Act (1933) 56, 102
bankruptcy 45, 46, 53
BARDA 98
Barofsky, Neil 63
Barrack, Thomas 89
Bebchuk, Lucian 50–52, 76, 77
Bell, Daniel 34
Benghazi attack 82
Better Markets 60
Bezos, Jeff 36
Biden, Joe 94
Binet intelligence test 64
biological theory 11
biology 9, 23
Biomedical Advanced Research and
  Development Authority 94
biostratification 26
birds 21. *See also* Pigeons
birth control 67, 68. *See also*
  abortion
  regulation 68
Black, Leon 46
black population
  female writers 17
blacks
  crime, accusations of 9
  cultural participation 7
  deportation to Africa 10
  economic discrimination 9
  eugenics 23, 68
  evolutionary differences 11
  genetic disadvantage 8
  genetic inferiority 9
  home buying, obstacles to 9
  human race, categorization of 11
  income gap 8, 9
  innate characteristics 9
  institutional discrimination 8
  intellectual deficiency 9
  intelligence with whites compared
    6
  IQ differences 7, 8
  land ownership 9
  New Deal, exclusion from 9
  perceptions of 11
  poverty 5
  reproduction, control of 68
  sexual characteristics 11
  socio-economic gap 5
  voting participation 65
  white elites, attitudes of 5
blue-collar workers 39, 91
Boeing 737 MAX disaster 52
Bok, Derek 44, 45, 50, 55
bonuses 41, 42, 45, 48, 50, 53, 63
  retention 63
Bradlee, Ben, Jr. 85
brain size 11
Bright, Rick 98
Brill, Steven 44, 54
  Tailspin 43
Brimelow, Peter 2
Brooks, David 86
Brown decision 9
Buchanan, Pat 12
Buckley, William F. 87, 89
Buffett, Warren 36
Burt, Cyril 28, 29, 31, 72
Bush, George W. 86, 101

C

capitalism 43
Carlson, Tucker 13–15, 92
Carlyle Group 46
Carnegie, Andrew 44
Carson, Ben 90
casino economy 40
Cattell, Raymond 65, 66, 68–70, 78, 79
Caucasoids 10
Center for WorkLife Law 91
Centers for Disease Control and Prevention 93
CEOs (Chief Executive Officers) 47–52, 58, 76, 107, 109
charity 107
Charlottesville Statement 73
Charlottesville torchlight rally 12, 32
children
 child custody 82, 96
 child support 34
 injury to 23
 middle-class 23
 minorities, outcomes for 23, 30
 poor academic performance 23
 poverty 30
 rights of 82
Chinese literature 7
Chomsky, Noam 39
citizen cooperation 14
civil liability 107
civil libertarianism 23
civil rights movement 9
class 4, 105. *See also* middle-class; working class
 structure 43, 88
climate crisis 35
Clinton Foundation 83
Clinton, Hillary Rodham 82–84, 90, 91, 101
cognitive ability 12, 13, 23, 64
cognitive elite 31–33, 38, 40, 42, 43, 45, 46, 50, 53, 54, 59–61, 64, 65, 68–70, 81, 87–89, 91, 94, 95, 107
 politics and intelligence 81–96
Cohn, Gary 57
Colasuonno, Lou 97
collective bargaining 104
Columbia Law 54
Commentary 6
communism 27
compensation 51, 107, 108
compensatory deprivation 30
compensatory education 34
computers 14
conservatism 32
consulting 42
consumers 55
 company contributions 50
 goods and services 47, 48
 products 54
 protection of 56
 rights of 56
contract buying 17
convict labor 9
coronavirus 53, 57, 92
 deaths, number of 93
 essential workers 107
 financial gains 107
 social control 92
 stimulus bill 56
 US Trump administration, response of 93–94
 vaccination 38, 74, 92
 White House Coronavirus Task Force 92

corporate lawyers 53–54
corporate management 47–60
corporate misbehavior 58
cost-effectiveness 31
Covid-19 pandemic. *See* coronavirus
credit default swaps (CDSs) 45
criminal behavior 59, 70
criminality
   organized anti-crime groups 14
criminal justice system 33
   homicide cases 14
Cruz, Ted 90
Crystal, Graef S. 50
Cuomo, Andrew 83
Cutten, George Barton 65

D

Damigo, Nathan 12
Darwin, Charles
Origin of Species 2
debt 36, 46
democracy 32, 33, 65, 66
   definition of 65
   economic 26
Democratic Party (US) 81, 84, 101
democratic socialism 26
Department of Justice (DOJ) 58
deportation
   of blacks to Africa 10
discrimination 5, 54
   discriminatory practices 9, 103
   inequitable treatment 8
   laws 8
   model 8
   socio-economic 23
distribution model 8
Dodd-Frank Act (2010) 56
Draper, Robert 89

Drucker, Peter 48
drug-related crime 15
Duke, David 12

E

Eckland, Bruce K. 72
economic growth 50, 59
economic inequality
   rise of 34
Economic Policy Institute 48
economic status 8
education 70
educational psychology 64
educational resources 30
efficiency principle 30
egalitarianism 25, 32
   egalitarian societies 27
eggheads 89
Eisenhower, Dwight 89
Eisinger, Jesse 58, 59, 77
electronic recordkeeping 14
Ellington, Duke 7
employment 70
   full 83
environmental factors 6, 15
environmentalist behaviorism 22
environmentalist doctrine 22
environmental protection 54
equality of opportunity 25, 27–30, 103, 105
equity 9
esthetic vandalism 87
ethnic differences
   in intelligence 4
ethnicities 7
ethnographic studies 61–62
eugenics 2, 23, 25, 66, 67, 69
   concept of 66

evolutionary selection 10
exceptionalist approach 104, 105, 107
experimental psychology 22

F

factor analysis 29
far-right extremism 12
fascism 2
   activists 12
Fauci, Lord 92
federal aid 31
Federal Deposit Insurance Corporation 56
Federal Register 54
fertility 68
FHA redlining 9
financial industry 41, 42, 44–46
financial reform 56
Financial Services Modernization Act (US) 102
fines 58
Fiorina, Carly 52
First Data 49
Fitzgerald, F. Scott 37
food stamps 106
Forbes 2
Forbes 400 36
Ford Motor Company 52
Fox News 92
franchise. *See* voting rights
Frank, R. 73
fraud
   bank 58
   fraudulent behavior 57, 58
   securities 58
   tax 89
free competition 27

Fried, Jesse 50–52, 76, 77
Frum, David 86
Fussell, Paul 37

G

Gage, Beverly 90
Gallup 44
Galton, Francis 25, 66
gang crime 15
Gardner, Howard 6
Gates, Bill 36
Gekko, Gordon 46
gender 105
general intelligence factor ("g") 29
General Motors 48, 49
genes 4–6
   role in human society 22
genetic aristocracy 24
genetic determinism 28
genetic enslavement 68
genetic inequality 32
genetic inferiority 9
genetics 6
   behavior 6, 23
   principles of 26
genotypes 23, 24
genthanasia 68
Gilded Age (nineteenth century) 36, 60
Gini coefficient 36, 37
Glass-Stegall. *See* Banking Act (1933)
globalized economy 85
goats 22
Goddard, H.H. 64, 65, 78
Goldman Sachs 42
Graham-Leach-Bliley Act (1999) 102

Graham, Martha 43
Great Compression (1940s–1970s) 34
Great Depression (1930s) 56
Great Divergence (1970s–present) 35
Great Recession 41, 45, 56, 58, 60, 83, 88, 102
Great Society programs 30
Green Party (US) 84
guns
  proliferation of 15

H

Hankins, F.H. 27
harassment 23
harmony 29
Harris, Kamala 83
Harris, Sam 3, 6
Harvard Educational Review 22, 23, 71, 82, 96
Head Start 34
health 2, 70, 106. *See also* coronavirus
  care management 40
  insurance 106, 107
  universal care 83, 106
hedge funds 41, 44, 45
Hemingway, Ernest 37, 74
hereditary factors 24
heritability 25
  concept of 23
  meaning of 24
Heritage Foundation 13
Herrnstein, Richard J. 1, 8, 10–12, 15–19, 21–28, 30–35, 38–40, 42, 43, 49, 53, 54, 59, 61, 69–74, 77, 79, 81, 82, 85, 95, 96, 99, 101
I.Q. and the Meritocracy 26
Hewlett Packard (HP) 52, 76
high-frequency trading 106
Hispanics 65
Hitler, Adolf 66
Ho, Karen 62, 63
home buying 9
  on contract 9
Home Depot 52
homicide 14, 15, 70
Hostess Brands 46
housing subsidies 106
human dignity 31, 69, 106
human efficiency 64
human labor 21
Hurd, Mark 52
hybridization 10
hypermeritocracy 37

I

Identity Evropa 12
immigration 65
  history of 10
  illegal 83
  policies 10
imprisonment 58
impulse control 11
income 2, 8, 34–38, 40, 41, 43, 44, 46, 47, 49, 51, 53, 54, 56, 57, 60–62, 64, 70, 71, 77, 85, 91, 95, 96, 102–106
Independence Party (US) 83
index scores 7
India
  caste system 65
Indian literature 7

individualism 33
inequality crisis 35
infant mortality 11, 70
inflation 35, 48
Ingle, Dwight J. 67
insider trading 58
Institute for Policy Studies 36
insurance 63, 105
   health 106, 107
intelligence 11
   alleged differences 5
   American level of 10
   biological differences 15
   concept of 24
   elite institutions as a marker of 62
   ethnic differences in 4
   genes and 21–23
   genetic 8, 15, 33, 38, 66
   Herrnstein's work 21, 23
   human value and 60–71
   measurement of 26, 29
   mediocre 28
   meritocracy and 30
   politics and 81–96
   racial differences in 6
   research 65
   spectrum 30, 39
   superior 38, 39
   testing 10, 22, 28
Intelligence Advisory Board 89
intelligence officials 89
interest rates 56
interracial relationships 10
investment banking 40, 61, 62
investment firms 62
IQ 4, 5
   class, relation to 15
   common household use of term 28
   datasets 34
   death depending on score 70
   distribution 31
   heritability of 25
   high scorers 69, 88
   high scorers 61
   homicide victims 70
   inherited factor 24
   low groups 32
   low housing 67
   occupations, list of 74
   racial differences 5, 7, 8
   report writing 14
   researchers 70
   scientists 66, 68
   scores 28, 29, 31, 34, 39, 61, 66, 69, 70, 103
   testing 10, 64, 65
Iraq
   military action in 83
Ivy League institutions 40

J

Japanese literature 7
J. C. Penny 53
Jefferson, Thomas 33
Jensen, Arthur 22–25, 30, 67, 71, 79
Jews
   World War II 9, 10
Johnson, Lyndon 23
Jones, Jacqueline 2
junk-bond market 57

K

Kasich, John 90
Kelleher, Dennis 60, 77

Klein, Ezra 3
Kristol, Irving 87
Krugman, Paul 92
Kubrick, Stanley
   Dr. Strangelove 69
Kushner, Jared 93

L

labor relations 56
land ownership 9
land rights 18
Laughlin, H.H. 10
law abidingness 11
law profession 54. *See also* corporate lawyers
legal chicanery 9
Lewis, Michael 90
libel 17
libelous articles 3
Libya
   military action in 83
Life Magazine 82
life, quality of 8
literacy levels 13
literature 8
   Asian 7
   Western 7
lobbyists 44, 53, 55, 56
Losing Ground
   American Social Policy, 1950–1980 4
Lowe's 52
low-skilled workers 107
Lynn, Richard 10, 11, 18

M

Madison, James 33, 81
malaria 68
management consulting 61, 62
Mankind Quarterly 9, 10
Manor Care 46
Margin Call (film) 40
Markovitz, Daniel 37, 41, 43, 49, 54, 57, 59, 74–78, 91, 98, 109
   The Meritocracy Trap 37
married couples 34
   unmarried mothers 34
Marxist theory 47
Marx, Karl 43
matching law 21
mathematics 7, 41
mayors 106
Mayors for a Guaranteed Income 106
MBAs (Masters in Business Administration) 42, 49
McDougall, William 65
McKinnell, Henry A. 52
McMillan Cottom, Tressie 105
Medicaid 106
Medicare 106
medicine 7, 40
mental abilities 24
mental health 11
mental testing 22, 28
mergers and acquisitions (M&As) 61
meritocracy 26, 30, 37, 42, 43, 72, 91, 101
   concept of 101
   hereditary 25
   IQ and 26
   meaning of 26
   meritocratic inequality 59
Merrill Lynch 40, 44
Metropolitan Museum of Art 87
middle-class 14, 23, 91

Milken, Michael 89
Mills, C. Wright 48, 53
  The Power Elite 48
minimum wage 83
monetary value 79
Morgan, J.P. 42, 44, 48
Morrison, Toni 8
mortgage-backed securities 45
Muilenburg, Dennis A. 52
Mulally, Alan 51
multiculturalism 86
multivariate methodology 65
murder 83. *See also* homicide
Murray, Charles 1–19, 21, 31–34, 38, 43, 69, 70, 73, 75, 79, 81, 82, 85, 86, 88, 94–96, 98, 99, 101, 102, 106–109
  Human Achievements 7
  index score 7
music 7
  genres 8

N
Nardelli, Robert 52
National Academy of Sciences 28, 47, 67
National Center for Immunization and Respiratory Diseases 94
National Economic Council 101
nationalism
  white 2
National Review 34
National Socialist government 12
natural selection 47
Nazism 66
  neo-Nazism 11, 32
  Rassenhygiene 9
  sympathizers 12

Third Reich 10
Negro 68
Negroids 10, 11
neoconservativism 6
neoliberalism 102–103, 105
New Deal 9
New Democracy 101
New Elite 96
New England Journal of Medicine 94
New Republic, The 1, 5, 6
New York Daily News 97
New York Post 97
New York Times 1, 2, 12, 40, 56, 70, 82, 86, 87
Niebuhr, Reinhold 60
  Moral Man and Immoral Society 60
Nixon, Richard 89
Nobel Prize in Literature 8
nuclear war 69

O
Obama, Barack 82, 90, 101
Obamacare 84
Occupational Safety and Health regulations 54–55
Oliver, Revilo P. 66
operant conditioning
  with Pigeons 21
Organization for Economic Cooperation and Development (OECD) 36
Orwell, George 68

P
Palin, Sarah 91
pandemic. *See* coronavirus

parenthood 67
parenting. *See also* children
  effort raising children 11
  quality of 2
  time and attention 11
Paulson, John 41, 89
PEN/Saul Bellow Award for Achievement in American Fiction 8
pensions 51, 52
perjury 58
personality 65
  traits 102, 103
Pfizer 52
philosophy 7
phones 14
Pigeons 22
  operant conditioning 21
  reconnaissance photographs 22
Piketty, Thomas 36, 74
  Capital in the Twenty-first Century 36
Pioneer Fund 9–12
Plato 27
  Platonic social model 33
  Platonic society 27, 28
  The Republic 27
podcasts 3
police
  effectiveness 14
  performance, deterioration of 14
Policy Review 13
political equality 81
politics
  intelligence and 81–96
population control
  selective. *See* eugenics
populism 84, 90
postgraduate qualifications 49

poverty 4, 36, 37, 39, 83
  federal antipoverty programs 4
  financial assistance 69
  Johnson's War on Poverty 23
  official assistance 15
  relative 39
pregnancy 67
presidential election (2016) 71
private equity 44, 46, 75, 89
privilege 85
proletarian labor 43
ProPublica 57
psychology 24, 65. *See also* educational psychology
public relations specialists 39
Pulitzer Prize 8, 58

Q

quality of life 106

R

race 4, 5, 9, 11, 13, 15, 16, 68, 69, 88, 105. *See also* blacks; whites
  American preoccupation with 4
  role of 4–16
racial differences 5, 8, 12, 24
  genetics of 13
  income 103
racial oppression 9
racial prejudice 10
racism 2–4
rape
  forcible 83
Rather, Dan 97
recession 41, 45. *See also* Great Recession
regulation rage 92

Reich, Robert 105
reproduction
   control of. *See* abortion; eugenics
reproductive strategy 11
   concept of 11
Republican Party (US) 82, 83, 96, 101
residuum 25
retail 40, 46, 53
Ritchie, David G. 27
Rivera, Lauren 62, 63
robbery 12, 66, 83
Roosevelt, Franklin D. 47
Roosevelt, Theodore 36
Rothwell, Jonathan 44
Rubalcava, Alex 61
Rubio, Marco 90
Rushton, J. Philippe 11, 18
Ryan, William 104

S

Saez, Emmanuel 35, 73
Sandel, Michael J. 105, 106
Sanders, Bernie 74, 88, 90, 91
SAT scores 40
scholarly brinkmanship 6
Schumer, Chuck 83
Schwartz, Nelson D. 38
   The Velvet Rope Economy 38
Scientific American 94
Scriven, Michael 68
segregation 9
selective deprivation 31
sensory tasks 22
sexual characteristics 11
sexuality 11
sexually transmitted diseases (STDs) 82

Shockley, William 67–69, 78, 79
sickle cell anemia 68
Silver, Nate 84
Simons, James 41
skillification
   of the US economy 105
Skinner, B.F. 21, 39, 42
slave labor
   of prisoners 9
slavery 27
Slobodian, Quinn 10
Snow, John 101
social cohesion 11
Social Darwinism 69
social democracy 103
socialism 27
social mobility 26
social policy (US)
   reform of 5
social psychology 103
social responsibility 26
social security 106
social standing 24
socioeconomic conditions 23
socioeconomic democratization 94
socioeconomic status 8
socioeconomic success 12
sociology 47, 48, 62–63, 105
Socrates 27
   myth of the metals 27
Spearman, Charles 29, 32, 33, 65, 66, 72, 78
Spencer, Herbert 69
Spencer, Richard 11, 32
Spitzer, Elliot 83
stealth compensation 51
Stelzer, I.M. 73
sterilization 66–68
   involuntary 67, 69

voluntary 67
Stevenson, Adlai 89
Stone, Oliver 46
student activism 23
subprime mortgage market 41
Sullivan, Andrew 6
Summers, Larry 101, 102
Sumner, William Graham 47
super-elite status 63
Supplemental Security Income 106
supremacism
   white 2, 3, 10–12

T

taxation 34
   tax code 56, 57
   tax havens 57
Taylor, Charles 86
Taylor, Jared 12
teaching 43
   teachers' salaries 43
technology 7
   advancement of 25
Terman, Lewis Madison 28, 29, 31, 65, 72, 78
Thatcher, Margaret 103
Thorndike, E.L. 47, 48, 64, 76, 78
Traditionalist Worker Party 12
Troubled Assets Relief Program 63
Trump, Donald 57, 82–98, 107
   presidential elections (2016) 83–85
Twinkies 46
twins 24

U

undocumented workers 83

unemployment 22, 25, 105
unions 34
   unionism, decline in 103–104
United States
   American caste system 34
   black immigrants 10
   collectivist 'American' identity 91
   discriminatory practices 103
   financialization of the economy 40
   financial professionals 44, 47
   Gini Coefficient, use of 36
   income hierarchy 36
   income inequality 36
   income level 41
   individualism, heritage of 33
   largest cities 106
   meritocracy in 27
   neo-Nazism 11, 66
   policy decisions, effect of 108
   political spectrum 101
   politics. *See* politics; Trump, Donald
   poverty rates 36
   Second Amendment 83
   skillification 105
   social policy 5
   unemployment rates 83
   universalist approach to inequality 106
   voting rights 65
   wealthiest individuals 36, 57
universal basic income (UBI) 106
universalist approach 104–106

V

venture capital 61
victim-blaming 104

Vietnam War 22, 82
Virginia Tech
  President of 6
vocation 43
  Latin definition 43
vocational education 29
vocational guidance 28
voting rights 65
Vox 3

W
wages 8, 37, 47, 107. *See also* income
Wall Street 17, 40–42, 44, 45, 53, 56, 61–64, 75, 76, 78, 84, 97, 98, 102, 108, 109
Wall Street (film) 46
Washington Police Department 13
Washington Post 13–15
Washington Summit Publishers 10
wealth
  addiction 42
  concentration of 35, 36
  distribution of 36
  inherited 36
  racial differences 103
Weiner, Anthony 83
welfare
  collective 43
  dependency 2
  disadvantaged persons 32
  national 32
  policies 67
  programs (US) 106
  public 60
  single women 106
  social 50
  state 102

Western culture
  superiority of 7
Western Michigan University 22
Weyher, Harry 12
whites
  cultural impact of 8
  discrimination against 5
  discriminatory practices 9
  fertility 68
  human race, categorization of 11
  income differences 8
  income gap 9
  intelligence with blacks compared 6
  IQ differences 7
  nationalism 2
  non-college-educated 86
  racist 4
  rednecks 5
  socio-economic gap 5
  supremacism 2
  white elites 5
  working class 85
white supremacism 3, 10–12
Will, George 87
Williams, Joan C. 91
Wilson, Charles 48, 49
woke society 85
women 17. *See also* abortion; sterilization
  black writers 8
  government assistance for low-income 34
  poor 69
  sexual assault against 82
  single 106
  smart women and birth rates 69, 70
  welfare for 106

workers
  safety of 54
  types of 29, 39
working class 35, 85
  neighborhoods 13
  superordinate 37
  traditional jobs 107
working-class 86
World War II 9
worthiness of human life 71

Y
Yale Law 54
Young, Michael 26, 29, 30, 72, 102
  The Rise of the Meritocracy 26

Z
Zucman, Gabriel 35, 73

Printed by Printforce, United Kingdom